HOME:现代时尚 高迪国际出版（香港）有限公司/编　胡小倩/译

HOME DÉCOR
Ideas for Interior Space, Function & Color

大连理工大学出版社
Dalian University of Technology Press

图书在版编目(CIP)数据

HOME：现代时尚：汉英对照/高迪国际出版（香港）有限公司编；胡小倩译. — 大连：大连理工大学出版社，2012.2
ISBN 978-7-5611-6347-4

Ⅰ.①H… Ⅱ.①高… ②胡… Ⅲ.①住宅—室内装饰设计—作品集—世界—现代 Ⅳ.①TU241

中国版本图书馆CIP数据核字（2011）第141040号

出版发行：大连理工大学出版社
　　　　　（地址：大连市软件园路80号　邮编：116023）
印　　刷：利丰雅高印刷（深圳）有限公司
幅面尺寸：246mm×290mm
印　　张：23
插　　页：4
出版时间：2012年2月第1版
印刷时间：2012年2月第1次印刷
责任编辑：刘　蓉
责任校对：李　雪
封面设计：四季设计

ISBN 978-7-5611-6347-4
定　　价：328.00元

电　话：0411-84708842
传　真：0411-84701466
邮　购：0411-84703636
E-mail：designbooks_dutp@yahoo.cn
URL：http://www.dutp.cn

如有质量问题请联系出版中心：（0411）84709246　84709043

HOME DÉCOR
Ideas for Interior Space, Function & Color

Preface | 序言

The world is being overtaken by technology; Ipods, Ipads, Kindles, Blackberries, Macs, Notebooks, Laptops, downloads' perhaps this is a good thing. Perhaps it's a sign of the inevitability of progress. But I don't think I'm the only one to regret the loss of the wild creativity of LP cover art, the pleasure of settling into a comfortable chair with a favourite book, or thumbing through the colourful pages of a magazine during the commuter rush hour. The thing all the items of inevitable progress lack is the pleasure of the tactile, of colour and design. And so, in a time where the new exists in a virtual realm, and equipment giving us access to this virtual world is of the stream-lined and soulless variety, we turn even more to our homes to fill what I believe is an innate human need for decoration, comfort and self-expression.

I don't think it's any accident that we have entered an era of "nesting". We're all working longer hours, travelling longer distances for both work and pleasure, and spending an increasing amount of time linked up to the virtual world via emails and social networking (and may I just ask why it is so necessary to be constantly contactable?). But we are human beings, and we can't ignore the 'being'. We like to touch, smell, see and feel. We are sensual creatures, and we take pleasure in things which make us feel good.

And there is immense pleasure to be had in a comfortable sofa, with room to stretch out your legs; in beautiful artwork, chosen because you love it, and which gives you delight whenever you look at it; in the vintage lamps which your partner found at the market in France during your holiday; at the Moroccan rug you lugged back from the weekend break to Tangier. These are all representations of the real you. No one puts these items of delight together quite like you. Your home is the canvas of your self expression, and like an artist you should be bold in your decoration. At home you are not judged; at home you are the king and queen of your castle. Why settle for the blandness of beige and the worry of "what is correct"? Trust your instincts. Bring together the things you love, whether they're antiques, vintage, contemporary or kitsch. Mix together orange and turquoise; blue and green. Do it a little at a time to start, if you're worried. Ask an interior designer for some guidance if you're not sure - the best interior designers will help you bring your inherent design preferences to life. They will help you create the rooms that are all about you, not all about them.

The pleasure of coming home to a place where you feel happy, where clutter is put away, where favourite treasures are displayed for your delight, where furniture is comfortable and well-placed, where you can put your feet up, this is what we all crave. When you are happy in your home, this is a big step to being happy in life.

ADRIENNE CHINN

科技正在引领着当今世界的发展：Ipod（苹果音乐播放器）、Ipad（苹果平板电脑）、电子书、黑莓手机、Mac技术、笔记本电脑、下载……，也许这是件好事，也许这象征着必然的发展。然而，我们失去了对LP封面艺术的疯狂创造力，失去了坐在舒适的椅子上阅读喜爱的书籍，或是在交通拥堵时翻阅杂志彩页的那份惬意。我并不是唯一一个对此感到遗憾的人。这种社会发展必然进程中的所有元素正是缺少了由触觉、色彩及设计带来的愉悦。因此，在不久的未来，当虚拟世界的新鲜事物，以及引领我们进入虚拟世界的设备源源不断涌现的时候，我们将把更多的目光投向我们居住的家，以满足我们所坚信的人类对于装饰、舒适和自我表达的固有诉求。

在我看来，我们已经进入了一个"巢居"时代，这并非意外。为了工作和娱乐，我们的工作时间越来越长，我们的旅行距离越来越远，通过电子邮件和社交网络，我们正花费越来越多的时间去链接虚拟世界（而我恰恰想问：为什么时刻保持联系变得如此重要？）。然而，我们是人类，我们无法忽视自身的存在。我们喜欢触摸，喜欢品息，喜欢观察，喜欢感知。我们是感觉的创造物，能够从一切美好事物中享受愉悦。

置身于舒适的沙发中，伸展双腿，徜徉于优美的艺术品中，选择您的最爱，无论何时欣赏它，都能给您带来快乐；在您爱人度假时从法国市场购买的古董灯饰下；在您周末休闲去丹吉尔买回的摩洛哥地毯上，所有这一切都将给您带来极大的乐趣和满足，为您展现一个真实的自我。没有别人能够像您一样，把这些钟爱的物品汇聚在一起。您的家正是您用来表达自我的一张画布，在装饰过程中，您应该像一个艺术家一样大胆创新。在您的家中，不需要由别人来评判；在您的家中，您就是城堡中的国王或是王后。为什么只满足于米色的柔和，而困扰于"何为正确"之中？请相信您的直觉。把您所钟爱的物品摆放在一起，无论它们是古董，还是当代艺术品，无论多么破旧不堪，亦或是粗制滥造。将橙色与蓝绿色，蓝色与绿色混搭。如果您会担心，那就从少许的尝试开始。如果您不能确定，那就请咨询室内设计师以获取指导。最优秀的设计师将给您带来最适合您的固有设计风格，为您打造一个只属于自己的专属空间。

回家的愉悦正是回到一个令您感觉幸福、远离纷扰的地方，一个依照您的喜好展示钟爱珍品的地方，一个家具舒适、摆放得当，并能翘起双脚的地方，而这正是我们所有的诉求。当您在家中感觉舒适畅快的时候，这便是迈向幸福人生的一个巨大跨越。

ADRIENNE CHINN

HOME DÉCOR
Ideas for Interior Space, Function & Color

Contents | 目录

006 | **Pavilion Park Show-Home**
亭子花园样板房

012 | **775 King Street West**
国王西街775号

022 | **Casa Son Vida**
颂维达别墅

034 | **Apartamento IS Lourdes**
卢尔德公寓

040 | **Apartment Chester Square**
切西特广场公寓

056 | **Orange Grove Residences**
橘林小区住宅

062 | **Apartamento Zaragoza**
萨拉戈萨公寓

070 | **Urban Chic**
优雅都市

080 | **Eduardo e Gina**
爱德华多·吉娜的家

088 | **15 Central Park West**
中央公园西街15号

094 | **33rd Street Residence**
33号大街公馆

106 | **Kensington House**
肯辛顿之家

114 | **Noho Loft**
诺赫顶层住宅

122 | **Georgian Townhouse**
乔治亚风格的联排别墅

128 | **Concrete House**
混凝土住宅

136 | **Winnington Close**
惠宁顿洋房

148 | **Kenig Residence**
肯尼格住宅

156 | **Cetatuia Loft**
塞塔图维亚阁楼

164 | **Collins Ave, Miami Beach**
迈阿密海滩的科林斯大街公寓

168 | **Vivienda 4**
维维埃达4号住宅

176 | **The Orange Grove**
新加坡中区橘林

184 | **Sentosa House**
圣淘沙岛住宅

196 | **Isabel**
伊莎贝尔

204 | **Riverside Drive**
河畔大道公寓

210	*The Lake by Yoo* Yoo湖区	296	*Buama* 布玛住宅
220	*Fractal Pad* 叠拼公寓	304	*HOME 07* 07住宅
226	*Pied a Terre* 临时住所	314	*The Woven Nest* 编织的巢
236	*The House Of Cesar, Roberta and Simone Micheli* 西萨、罗伯特和西蒙·米歇利的家	318	*Greenwich Village Penthouse* 格林威治村顶层公寓
244	*Walmer Loft* 瓦尔默阁楼	328	*Shelter Island House* 牛尾洲住宅
252	*White Apartment* 白色公寓	334	*Attico* 阁楼
258	*Absolutely Fabulous* 美妙绝伦	338	*Penthouse Serrano* 塞拉诺顶层公寓
264	*56th Street Apartment* 56号大街公寓	346	*Twin Loft* 双子座阁楼
270	*Murray Hill Townhouse* 默里山联排别墅	352	*Istanbul Suites* 伊斯坦布尔套房
276	*Setia Eco Villa* 塞提亚生态花园别墅	360	*35 White Street* 怀特街35号
280	*Tite Street* 泰特街	364	*Index* 索引
288	*Ojeni Flats* 欧杰尼公寓		

HOME DÉCOR
Ideas for Interior Space, Function & Color

Pavilion Park Show-Home
亭子花园样板房

One of the unique aspects of this Show-home was that it had an internal light-well throughout the core of the building. Some potential buyers from previous viewings (of the unfurnished space) had expressed some concerns regarding the practical maintenance of the light-well.

CWD's brief was to dress the interior space and help to position the light well as an asset to the potential buyer's lifestyle, rather than be a perceived as a maintenance drawback.

These Terrace homes in Bukit Batok Singapore, were a mid-end product for the client and was targeted towards Local Singaporeans that were wanting to updated their homes from public housing to private landed housing, which is what many Singaporeans aspire to.

CWD decided that in order to target these group of 'Upgraders' it would not be appropriate to creat a ultra high end interior as this would be deemed unaffordable by many of the target group, instead we decided to create an interior designed but in a way that was not too complex that potential buyers could emulate elements if they chose to do so themselves.

The designer's analogy was, instead of designing the Calvin Klein couture collection equivalent for the show-home, we would create the CK Collection, providing entry level access into the world of interior design, perfect for this mid tier, upgraders market.

Location: Singapore
Area: 265m²
Designer: Cameron Woo
Photographer: Masano Kawana

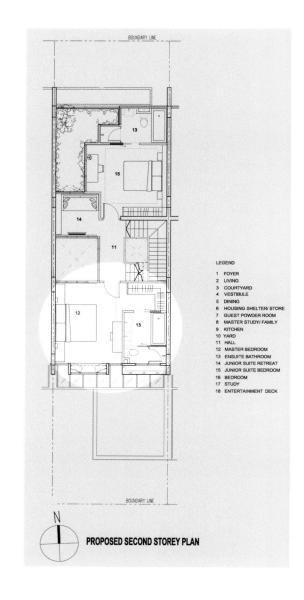

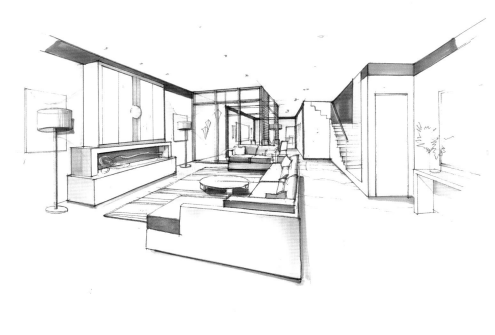

这套样板房的独特处之一在于一个内部天井贯穿了整个建筑的核心区域。一些潜在的客户通过事先（对不带家具的清水房）观察，对其采光的实际效果表示担忧。

设计师卡梅伦·吴的任务是要对样板房的内部空间进行装饰，帮助潜在客户实现良好的采光效果，而不是掩盖其缺陷。

在新加坡的武吉巴督，这些连栋房屋是为客户推出的中端产品，目标定位于新加坡本地客户，他们希望从公共住宅升级为私人住宅，这是他们梦寐以求的。

为了锁定这一"升级型"客户群体，设计师卡梅伦·吴决定呈现一种不太复杂的室内设计方案，那些潜在的客户可以自行选择仿效设计元素。因为超高端的室内设计会令目标群体中的很多人无法承担。

设计师不是要设计相当于卡尔文高级时装系列的样板房，而是希望创造出卡尔文系列，针对中层客户和寻求改善居住条件的消费者，提供世界品牌入门级的室内设计。

HOME DÉCOR
Ideas for Interior Space, Function & Color

775 King Street West
国王西街775号

The quality of Toronto's King West' suites are demonstrated in the two spectacular models designed by II BY IV. Both models include a brick feature wall to reflect the neighbourhood context. In the two-bedroom unit, this harmonizes exceptionally well with the natural tone palette and emphasis on texture the designers have employed. Black ash laminate flooring is topped with a taupe carpet with a tree graphic in the living room, grounding a seating group in felt and textured wool accented with worsted flannel cushions. The den features another graphic carpet below the smart grey felt daybed. In the kitchen, the soft grey tones are picked up in the mobile breakfast counter and the whitewashed grey cabinetry above a backsplash of back-painted glass and marble. Black stone counter tops and streamlined hardware complete the understated style.

In contrast, colours pop against the one-bedroom unit's white ash laminate floor, white walls and obscured glass partitions. Simple white kitchen cabinets contrast with a poppy-coloured glass mosaic backsplash, clever cantilevered wooden dining surface and matching bench. In the loft-style living room, a colourful super graphic portrait, and pop-art cushions add brightness and fun to a collection of quirky furnishings from a woven wool ottoman, raw Douglas firm coffee table, and pale walnut mid-century TV console to the transparent acrylic accent pieces.

With brick feature walls in common, these two suites reflect and attract two lifestyles: sophisticated professionals as well as youthful buyers choosing to make their first home close to nightlife, cultural and sports attractions. Both suites contain many natural references, yet one is an urbane mix of subdued tones with rich textures, while the other is a lively melange of pop art and bright colour.

Location: Toronto, Canada
Designer: II BY IV Design Associates

013

　　这两个惊艳绝伦的设计由II BY IV设计公司完成，彰显了多伦多国王西街套房的品质。两个样板间都包括一面砖砌的装饰墙，与周围的景致交相辉映。在两居室的单元里，这面装饰墙与自然的色调格外和谐，并且着重突出了设计师选用的材料的质地。客厅，黑桦木强化复合地板，上面铺有树形图案的灰褐色地毯，沙发座椅包裹了带纹理的羊毛面料，搭配精纺法兰绒靠垫。小房间则在灰色毛毡两用长椅下铺设了其他图案的地毯。厨房以柔和的灰色调为背景，设有可移动式早餐台，背面涂色的玻璃和大理石后挡板上安装了白色的细木板橱柜。黑石台面以及一字排开的硬件设施展现了一种低调朴素的风格。

　　相反，在一居室的单元里，白桦木强化复合地板、白色墙面，以及毛面玻璃隔板，一些颜色在这样的背景下格外抢眼。简洁的白色橱柜与深红色的玻璃马赛克后挡板、精巧的悬臂式木质餐台和配套的椅子形成强烈反差。在阁楼式的客厅里，一幅色彩艳丽的精美图画和流行的艺术靠垫为一系列奇特的陈设增添了活力与情趣，比如，土耳其羊毛织物、原始的道格拉斯咖啡桌、浅色的中世纪胡桃木电视柜，以及透明的亚克力装饰物。

　　虽然设计了同样的砖砌装饰墙，但两套样板间反映并吸引了两种生活方式，即资深的专业人士以及年轻人，一些年轻客户希望自己的第一个家能够靠近文化和体育中心，能够享受丰富的夜生活。两套设计均吸纳了许多自然元素，一个是融合了平缓基调与丰富质感的温文尔雅，而另一个则是流行艺术与亮丽色彩的生动聚合。

HOME DÉCOR
Ideas for Interior Space, Function & Color

Casa Son Vida
颂维达别墅

Marcel Wanders created the interior design for Casa Son Vida on the island of Mallorca, a luxury villa of contrasts. The building that houses Casa Son Vida is composed of an old and a new part. The stunning architecture (by tecARCHITECTURE) of the new extension inspired Marcel Wanders to complement the building with a highly exclusive, captivating interior, linking old and new. The round and square shapes, soft blobs and new antiques create a unity of the architecture and the interior design. A mix of traditional and modern references is visible throughout the villa, from the classic profiled wall lining the curved space to the newly custom designed cupboards in straight lines. All aspects of the space and dimensions are played with by using relief and contrasting smooth surfaces, creating a unique atmosphere.

Location: Mallorca, Spain
Area: 790m²
Architect: tecARCHITECTURE
Interior design: Marcel Wanders Studio
Developer: Cosmopolitan Estates
Photographers: Gaelle Le Boulicault, Marcel Wanders

设计师Marcel Wanders对位于马略卡岛的颂维达别墅进行了室内设计，展现了一种富于对比的奢华。该建筑由新旧两部分组成。tecARCHITECTURE公司对原有建筑进行了全新的扩展。富于魅力的建筑风格给予Marcel Wanders设计灵感，他运用极为奢华、抢眼的室内装饰将传统与现代完美结合。

这些或圆或方的造型、柔和细腻的颜料、新奇的古董工艺品使建筑本身与其内部装饰浑然一体。从典雅的异型墙体所塑造的弯曲空间到一字排开的全新定制设计的橱柜，在这套别墅里，随处可见一种传统与现代的混搭。通过运用浮雕与平面的对比，营造了整个空间的独特氛围。

HOME DÉCOR
Ideas for Interior Space, Function & Color

Apartamento IS Lourdes

卢尔德公寓

The predominant idea of this project, signed by Ulisses Morato, is the architecture of the essential, i.e., the one associated to a minimum intervention with the maximum visual and functional impacts. The functionality here is translated into ample spaces with generous circulation, reduced furniture with rational distribution and materials that facilitate daily maintenance of the apartment.

The comfort was guaranteed by the excellent ergonomics of the furniture, the mildness of colors, the efficient artificial illumination, associated to the good conditions of natural illumination and ventilation.

Ornaments and works of art gain visual strength due to the large empty spaces that surround them and to the predominant white background.

The revitalization of the apartment comprised the architectonic and decorative elements into a single project: re-structuration of layout, design of furniture, specification of ornaments, Persian blinds, curtains, lamps and finishing materials. This process was important to guarantee the final balance among all elements that compose the environments. The floor in marble received a new polishing, the walls gained new colors and the ceiling was lowered with cardboard plaster plates in order to receive new illumination. The lights mark the circulation axles, distinguish the works of art and ornaments, and allow the residents to have innumerable scenarios to be used by them.

Location: Minas Gerais, Brazil
Area: 250m²
Designer: Ulisses Morato
Design company: Morato Arquitetura
Photographer: Daniel Mansur

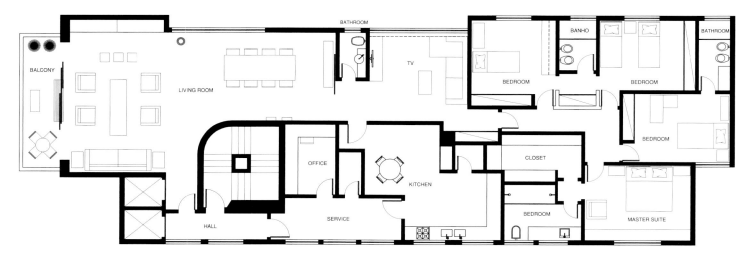

FLOOR PLAN

　　尤利西斯·莫拉托签订的该项目，其主导思想在于建筑的精髓，将最小的干预与最大的视觉效果和功能性整合到一起。设计的功能性体现在流通性好的宽敞空间、通过合理布局实现的家具简化，以及易于日常维护保养的装饰材料等方面。

　　家具出色的人性化设计、柔和的色调、有效的人工照明，以及优越的自然照明和通风条件，这些因素保证了设计的舒适性。

　　装饰品和艺术品在宽敞的空间和白色背景的映衬下，形成强烈的视觉冲击。

　　这个改造项目将建筑艺术与装饰元素融为一体，包括布局的改造、家具的设计、装饰的品质、百叶窗、窗帘、灯具和其他饰面材料。整个设计过程对于保证这些元素之间的最终平衡起到了至关重要的作用。设计师将大理石地面进行重新抛光，重新喷涂墙面，利用硬石膏板降低天花板的高度以便改善照明效果。灯具的位置标示了照明设计的中轴，令房间的艺术品和装饰品更引人瞩目，住户可以在此尽情享受生活的惬意。

HOME DÉCOR
Ideas for Interior Space, Function & Color

Apartment Chester Square

切西特广场公寓

Located in Chester Square, this apartment is designed by Helene Benhamou, who has been engaged in the world of art with her own unique elegant style for more than 20 years. The apartment is comprised of dining room, sitting areas, study, kitchen, TV room, music room, bar, master room and three rooms for children. A warm atmosphere is created by an aluminum 'LOVE' sculpture in dining room and a 'FIVE' in sitting room. Helene's tastes for design are evident in sofa and the old-fashioned wooden armchair she chose for the house.

Location: UK
Designer: Helene Benhamou
Design company: HB Interior

这个位于切西特广场的公寓是由室内设计师Helene Benhamou所设计的,在艺术的世界中徜徉超过20年的她有着自己独特的优雅的设计风格。此公寓包括餐厅、客厅、书房、厨房、电视室、音乐室、酒吧、主卧和三个儿童房。餐厅里铝制的"LOVE"雕塑让餐厅的气氛更加温馨,客厅里的"FIVE"雕塑也增添了情调。公寓里的沙发、古旧的太师椅等也都是设计师精心挑选的,设计师的品位也从中得到了完全的体现。

HOME DÉCOR
Ideas for Interior Space, Function & Color

Orange Grove Residences
橘林小区住宅

Cameron Woo Design's brief was to create a sophisticated and elegant interior that would attract affluent owner-occupiers or foreign investors. The potential buyers would most likely be successful and wealthy families, from India, Indonesia or China. In order to communicate the design concept for the OGR penthouse, CWD decided to choose a celebrity that would best represent the 'HOLLYWOOD GLAMOUR' lifestyle that we were trying to create, in order to attract the desired target market. CWD chose Aishwarya Rai as the International celebrity (famous for winning the Miss World crown in 1994), whose lifestyle we would model and emulate, for the Penthouse Interior. Ms Rai is a successful actress and Bollywood superstar whose glamour, elegance, exotic background, appreciation for colour (reflected in her fashion wardrobe) and celebrity lifestyle, would reflect the same celebrity qualities in the Penthouse suite at Orange Grove Residences. CWD felt colour was important to differentiate from the other zen inspired show-houses in the nearby vicinity, and since Ms Rai was not afraid to wear vibrant colours in her clothing, the designer hypothesized this would also be translated in her home interior as well.

Location: Singapore
Area: 341m²
Designer: Cameron Woo
Design company: Cameron Woo Design

LOWER FLOOR PLAN

UPPER FLOOR PLAN

设计师卡梅伦·吴力图向人们展示一个精致高雅的室内设计，吸引那些高端客户或是海外投资商。该住宅区的潜在客户大多来自印度、印度尼西亚和中国的成功人士及富裕家庭。为了传达橘林小区顶层公寓的设计概念，卡梅伦决定选择一个最能代表我们一直努力营造的"好莱坞奢华"生活方式的社会名流做代言，以便吸引我们所期望的目标市场。

最终，设计师卡梅伦·吴选择了艾西瓦娅·蕾（1994年，她因荣获世界小姐桂冠而声名鹊起），她的生活方式正是我们这栋公寓内部设计希望塑造并效仿的。艾西瓦娅·蕾是一位成功的宝莱坞女星，她的优雅魅力、异域风情、对色彩的品位（反映在她对时尚服饰的选择上），以及名媛的生活方式，都反映出该住宅区业主的生活品位。卡梅伦·吴发现，要想令该设计在附近其他以禅宗为设计灵感的样板间中脱颖而出，颜色起到了至关重要的作用。既然艾西瓦娅·蕾女士都不惧怕穿着鲜亮颜色的服饰，设计师猜想这一理念也会融入到她的室内设计中。

HOME DÉCOR
Ideas for Interior Space, Function & Color

Apartamento Zaragoza
萨拉戈萨公寓

An integral retrofit was planned, with a complete renovation as main target, modifying all elements in the house, such as distribution, light, design, facilities, etc. The result was completely satisfactory to the client. The apartment has its access through the upper floor, with an entry that defines the concept and style of all other spaces. Only few paintings with sounded shapes and grey and blue colours, painted by the artist Mercedes Rodriguez, along with a grey carpet and several steel iron balls by 'A-cero in' break the white uniformity. An opening in the ceiling, let the light in through an existing skylight. This entrance gives access to the more public rooms, composed of an spacious living room at one side, and a toilet, a bedroom en suite, both for guests or services and the kitchen at the other side. Opposite the entrance, the sculptural staircase gives us access to the more private rooms. At the living room, as well as in all other spaces, the white colour has the leading role. Just few elements as the black lacquered wood console table, the sofas, Ace model by 'A-cero in', with champagne colour velvet stand out on the white monotony. At the other side, in the same floor, the staff accommodation is located, with a toilet, a kitchen and a staff or guests bedroom en suite. The kitchen with a longitudinal layout was specially designed by A-cero for this space, where the whiteness brings out the black table made in Hi-Macs.

The project's objective was to make the most of its space, optimizing the distribution of the apartment, at same time the design and the light are emphasized. As light is the main target in this project, the white colour both in ceilings, walls and flooring with a white special lacquered wood, combined with the black colour to emphasize different spaces, creating sculptural volumes.

Location: Saragossa, Spain
Area: 250m²
Design company: A-cero Joaquin Torres architects
Photographer: Luis H. Segovia

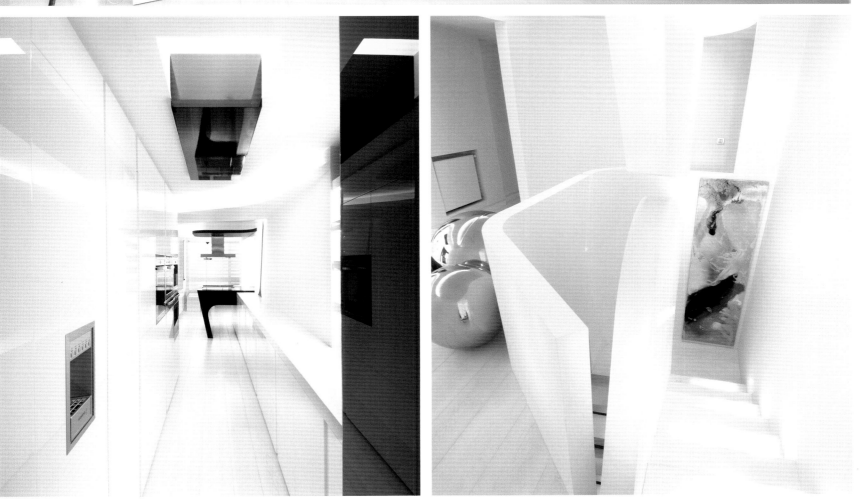

这是一个整体改造项目，以彻底翻新为主要目标，改变公寓内的所有元素，比如，布局、灯光、设计、陈设等。客户对该设计的成果十分满意。

公寓内设有楼梯通道，入口处诠释了所有其他空间的概念和风格。几幅造型优美、蓝灰相间的绘画出自艺术家梅赛德斯·罗德里格斯之手，搭配灰色的地毯和几个刻有"A-cero建筑师事务所"的钢铁球，打破了白色的统一性。自然光透过屋顶天窗直射屋内。这个入口可以通向多个公共区域，包括位于一侧的宽敞的客厅、洗漱间、供客人和佣人使用的卧室套间，以及另一侧的厨房。入口对面，雕刻造型的楼梯通向更多私密空间。客厅和其他所有空间均以白色为主色调。只有少数元素从单调的白色中脱颖而出，比如木质的黑色喷漆边桌、沙发、刻有"A-cero建筑师事务所"的模型，以及香槟色天鹅绒。在同一楼层的另一侧是员工宿舍，设有洗漱间、厨房、供员工或客人使用的卧室套房。纵向布局的厨房是设计师专门为此量身打造的，白色背景令Hi-Macs黑色餐桌尤为醒目。

该项目旨在尽可能地利用空间，优化公寓布局，同时强调设计和灯光。由于灯光是该项目的设计重点，因此，天花板、墙面和地板也使用了特制的白色油漆木材，结合黑色点缀来凸显不同的空间，营造出一个雕塑般的立体空间。

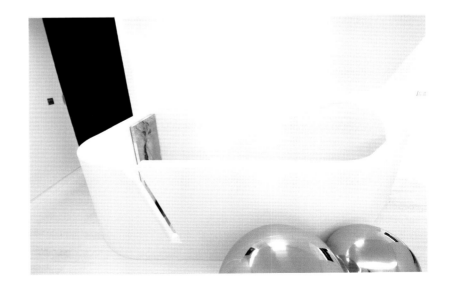

HOME DÉCOR
Ideas for Interior Space, Function & Color

Urban Chic
优雅都市

The interiors were inspired by the urban London environment - this seemed natural as you can see many iconic London landmarks from the balcony, including the London Eye, Battersea Power Station, and Big Ben. The client wanted a very modern interior, so the color palette chosen was tones of grey, white, red and black. The choice of grey, white, red and black was very much inspired by London itself. London is rainy, so the grey skies are a normal feature of London life. White comes from the clouds, red is from the bright red buses and post boxes, and black comes from the famous London taxis.

All the artwork is by British artists and two rugs from Adrienne Chinn's Urban London rug collection were custom-made for the living room and dining room. The rug designs were inspired by photos Adrienne took on the streets of London - the red graffiti on the dining room rug was on the top of a bus shelter and the pattern in the living room rug comes from marks on the pavement on the King's Road in Chelsea. Grey polished plaster accent walls in the dining room and hallway were inspired by London concrete, as were the textural feature walls in the bedrooms - the wallpapers are made up of thousands of chips of stone which reflect the light beautifully!

Location: London, UK
Designer: Adrienne Chinn
Design company: Adrienne Chinn Design
Photographer: James Balston

　　该室内设计项目的灵感源自伦敦的城市环境，正如您从阳台所看到的许多伦敦的标志性建筑，包括伦敦眼、巴特西电站和大本钟一样，这里也透露着一份自然的淳朴。客户希望呈现的是现代室内设计，所以选取了以灰、白、红、黑为主的色调。色彩选取的灵感同样来自于这座城市。伦敦常年多雨，因此，灰色的天空是这里最主要的特征。白色是云彩的颜色，而红色则是公交车和邮筒的颜色，黑色代表了著名的伦敦出租车。

　　所有的艺术品均出自英国艺术设计师之手，客厅和餐厅的两块地毯是从Adrienne Chinn的伦敦地毯艺术藏品中心定做的。地毯的设计灵感来自Adrienne在伦敦大街上拍的照片——餐厅地毯上的红色涂鸦模仿了公共汽车站候车亭的顶棚；而客厅里的则是切尔西国王大街便道上的路标。餐厅和玄关，灰色光面石膏主题墙的设计源自伦敦的混凝土建筑。卧室的纹理主题墙贴有无数碎石制成的墙壁纸，在灯光的映衬下，分外妩媚。

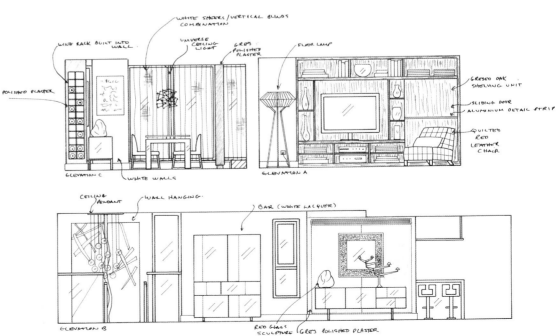

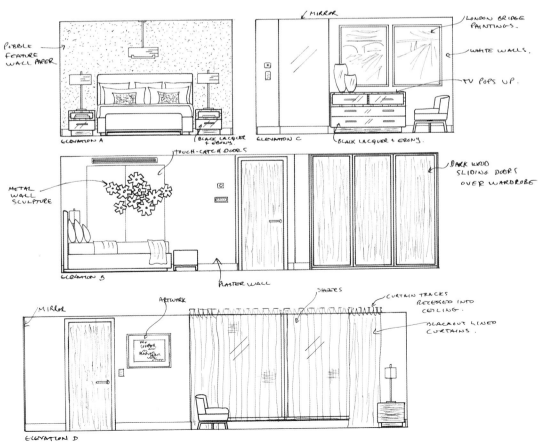

HOME DÉCOR
Ideas for Interior Space, Function & Color

Eduardo e Gina
爱德华多·吉娜的家

The project is a 350m² apartment, for a couple and their three sons. The requirements were: amplitude, integration with space for being together with family and friends.

In the kitchen, the architect abused of the black, creating a sober room, modern and elegant, with 17m².

The white floor, with big boards with 80x80cm is from Porcelanato High Gloss Peace, da Recsa. The walls are coated by Porcelanato Cetim Bianco, from Portobello. Cabinet are Florense and matches white with High Gloss black in the doors. Handles shows in a discrete aluminium form. The superior closer from the cabinets and the door are in black glass, executed by Penha Vidros. The central stand is all made by black High Gloss.

The imported equipments are from the special line of Viking. The covers are made by marble, Nero Mchurra. The special gourmet vats are from Metald. The chef mixer, which brings the practicality to the kitchen is from Punto. And, finally, lightning was made by La Lampe.

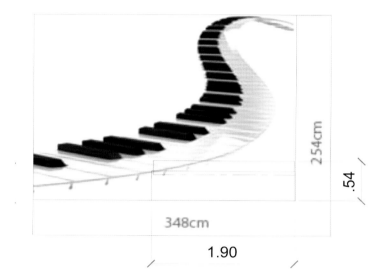

Location: São Paulo - Jardim Anália Franco
Area: 350m²
Designer: Brunete Fraccaroli

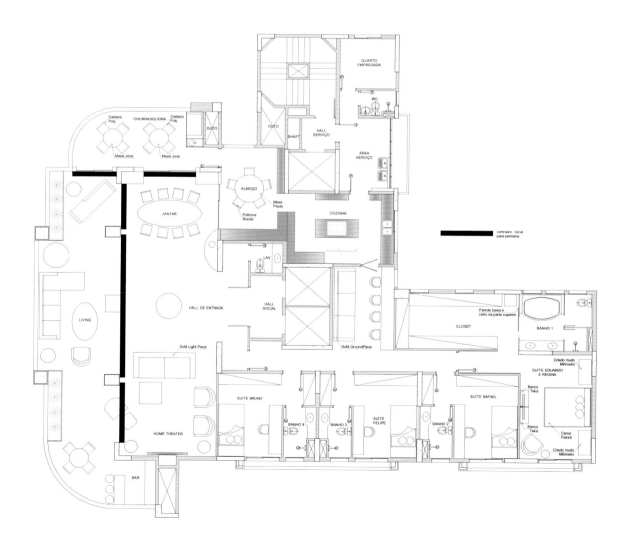

该项目是一栋350平方米的公寓，主人是一对夫妇和三个儿子。要求设计拥有足够宽敞的空间供家人和朋友聚会。

17平方米的厨房，设计师运用黑色营造出一个既清宁高雅，又时尚现代的空间。

地面铺设的是80厘米见方的白色大块板材，出自Porcelanato High Gloss Peace, da Recsa。墙面是产自爱丁堡波多贝罗的Porcelanato Cetim Bianco。Florense的橱柜，白色柜体配有黑色高光门板，铝制把手镶嵌其上。柜体上的superior closer和门均使用了Penha Vidros的黑色玻璃。中间的岛台全部采用黑色高光板材。

一些进口设备是从Viking电话购物买来的。厨房的台面是Nero Mchurra大理石。专用厨具来自Metal，非常实用的搅拌机是Punto的。最后，还有La Lampe的照明系统。

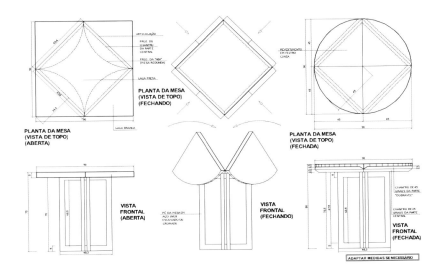

HOME DÉCOR
Ideas for Interior Space, Function & Color

15 Central Park West
中央公园西街15号

The client came to D'Aquino Monaco Inc with the desire for 'style and color' for their new apartment on Central Park West. They were seeking a 'varied experience with memorable rooms.' By eliminating service corridors, and rearranging the kitchen, we created a spacious backdrop for a new collection of modern art and fine furnishings. The project became an artistic dialogue with the clients as well as other artisans. The foyer moldings were collaboration with Moss and Lam and the D'Aquino Monaco office to establish a sculptural identity for the entrance, removing all traces of the original prewar style crown molding. This set the tone and style for the apartment, the blending in 20th century Italian and French antiques with modern sculptural artworks. Art by Richard Artschwager intended for the wall became a ceiling piece in the Family Room; it is balanced by a gravity defying storage wall on chrome springs designed by D'Aquino Monaco. In a similar manner a sensuous Lucite sculpture by Jim Shaw becomes a unique coffee table in the living room. Love for all things 1940's, particularly Italian, is expressed in the extensive use of Fornasetti wallpaper lining the kitchen, breakfast room, and laundry. Although each room is unique and has its own identity, the apartment is brought together by whimsical dialogue between art and design.

Location: New York, USA
Designer: Carl D'Aquino, Francine Monaco
Design company: D'Aquino Monaco Inc.
Photographer: Peter Margonelli

客户来到D'Aquino Monaco设计公司，希望为他们在中央公园西街的新公寓找到理想的"设计风格和色彩"。他们正在寻找的是一种"随着那些令人难忘的空间而变化的生活体验"。设计师通过拆除现有的走廊并重新布置厨房，为一系列崭新的现代艺术品和精美家具打造出一个宽敞的背景。设计将这里变成了一个与主人和其他艺术家对话的艺术空间。拆除了门厅原有的战前原始风格的天花板棚线，取而代之以雕塑般的造型，由Moss、Lam和D'Aquino Monaco设计公司共同设计完成，这个造型融合了20世纪意大利和法国古董及现代雕刻艺术品，它奠定了整个公寓的设计基调和风格。设计师将理查德·阿茨希瓦格设计的墙面装饰艺术品挪到了家庭娱乐室的天花板上，同时，利用D'Aquino Monaco设计的铬合金弹簧来保持反重力壁橱的平衡。类似的设计还有客厅的独特的咖啡桌，出自吉姆·肖之手，采用了富于美感的透明合成树脂雕刻技术。整个设计通过在厨房、早餐间和洗衣房内大量使用的Fornasetti壁纸，表达了对20世纪40年代的所有艺术品，尤其是意大利风格的钟爱。尽管每间房都是独一无二且别具风格，但整栋公寓还是在艺术与设计之间展开了一场异想天开的对话。

HOME DÉCOR
Ideas for Interior Space, Function & Color

33rd Street Residence
33号大街公馆

Located off a walk street in Manhattan Beach with panoramic ocean views, this 418 square meters residence was designed as a modern South Pacific Island retreat.

Drawing inspiration from their favorite vacation destinations, the clients set the tone for their ideal lifestyle. The intent was to achieve an 'architecture pure, clear, clean, neat and healthy,' possessing tropical overtones, natural materials and a hint of luxury.

Organized efficiently, like a ship the home design parallels moves often seen in boat architecture. A nautical theme runs throughout, from the expansive and free 'upper deck' (third level) where living is open, light and breezy, to the tidy organization of the sleeping quarters 'below deck' at the mid-level. The central stair at the heart of this home with its wood floating cantilevered teak planks accompanied by a mahogany slatted ladder, appears as a sail-like screen soaring nearly three stories tall.

From its organization and structure to its refined details and tropical finishes, this modern South Pacific Island retreat feels right at home at the beach.

Location: Manhattan Beach, California, USA
Area: 418m²
Architect: Christopher Kempel, AIA, Rocky Rockefeller, AIA
Firm: Rockefeller Partners Architects
Landscape Architect: John Feldman
Interior Designer: Alana Homesley
Photographer: Eric Staudenmaier

该项目位于曼哈顿海滩的一条步行街上，面积为418平方米，可以饱览海滩全景，是一栋现代的南太平洋海岛住宅。

房主从他们最钟情的度假胜地汲取灵感，确立了一种理想的生活方式。设计目的是要表现"纯净、明亮、整洁和健康"这一主题。彰显热带风情，既天然环保，又不失奢华魅力。

房子设计的有序布局效仿了船舶建筑设计的平行移动理念。航海主题贯穿始终，从宽敞自由的"上层甲板"一直延伸到中间夹层整洁有序的卧房。"上层甲板"位于第三层，作为开放的起居空间，光线充足且通风良好。房子中央的楼梯由悬臂式柚木支架和红木踏板组成，三层高的设计宛如扬帆起航的船只。

从布局到结构，从精致的细节到热带风格的修饰，这个现代的南太平洋休闲寓所定会让人有宾至如归的感觉。

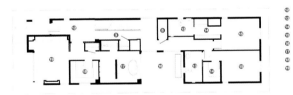

12 entry
13 bath
14 closet
15 bedroom
16 zen garden
17 master bath
18 master closet
19 master bedroom

ground
1,674 square feet

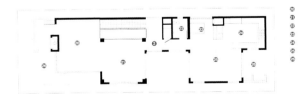

20 living room
21 dining room
22 powder
23 breakfast nook
24 kitchen
25 family room
26 outdoor deck

upper
1,330 square feet

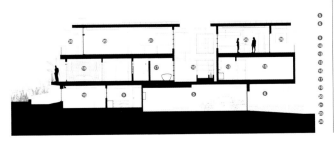

5 three car garage
6 two car outdoor covered parking
9 study
10 beach room
13 bath
15 bedroom
16 zen garden
17 master bath
19 master bedroom
20 living room
21 dining room
25 family room
26 outdoor deck

n/s section

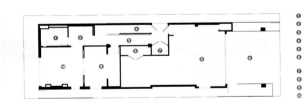

1 beach bath
2 beach bar
3 hall
4 storage
5 three car garage
6 two car outdoor covered parking
7 laundry
8 mechanical
9 study
10 beach room

basement
1,496 square feet

4,500 total sf

site

HOME DÉCOR
Ideas for Interior Space, Function & Color

Kensington House
肯辛顿之家

British architects and designers SHH took on the new-build Kensington House project with a commission to create a complete interior scheme for the six-storey, 1,114 sqm house. The new-build house is a contemporary take on a Georgian terrace house, set within a classic Square, combining all the conveniences of modern living with state-of-the-art AV and comfort cooling, without compromising on space and traditional proportions.

The property is comprised of a very generously proportioned reception room, large formal dining room, family room, cinema and pool room, all located on the lower 3 floors. The entire first floor is occupied by the master suite, with the remaining two floors taken up by 5 further bedrooms and en suite bathrooms.

The client's brief was simple. The house had to suit the needs of an international couple with a large family.

Location: London, UK
Area: 1114m²
Designer: Rene Dekker and team at SHH
Design company: SHH

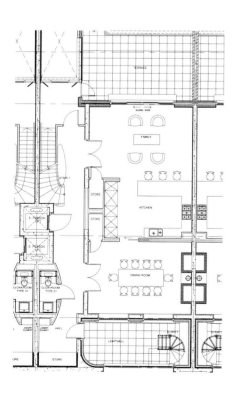
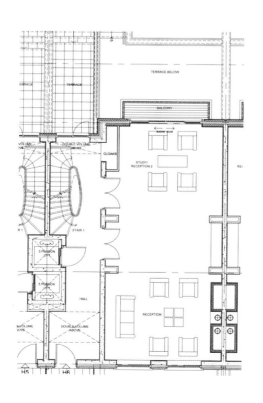
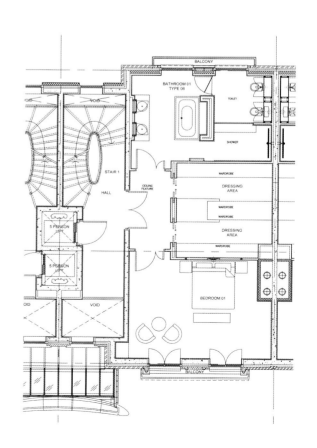

这个名为"肯辛顿之家"的六层楼全新整体室内设计项目总面积为1114平方米，由英国建筑师兼室内设计专家SHH亲手打造。新建筑位于一个古典风格的广场上，具有乔治王朝时代的特色，却不乏现代气息。这里具备了所有现代生活必需的便利条件、最先进的视听系统以及舒适的冷气，而这些也并没有影响空间及传统的设计比例。

整栋建筑为设计留有极大的施展空间。一至三层设有接待室、宽敞的正式餐厅、家庭娱乐室、电影放映室，以及台球室。此外，一楼整层是主人的套房，剩余的两层楼还有五间配有浴室的卧室。

客户的要求很简单。房子必须符合一对生活在大家庭里的国际夫妇的需求。

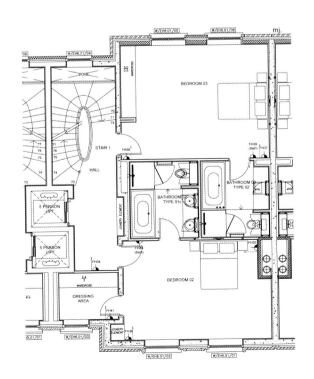
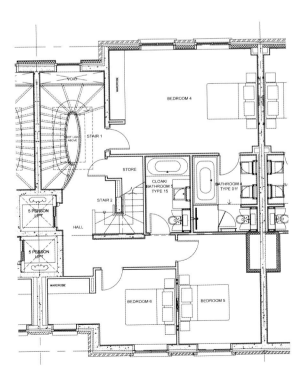

HOME DÉCOR
Ideas for Interior Space, Function & Color

Noho Loft
诺赫顶层住宅

The design of the loft is in a way antithetical to Manhattan's Noho neighborhood, where the common practice is the open plan. The approach for this residence was to create gallery-like vistas in the living areas and separate the sleeping spaces to allow each its own identity and privacy.

The children's wall provides a home within the home, combining storage for toys, an Astroturf covered play area, a lofted bunk, and nestled, intimate "cocoons" for leisure and reading. Situated to the scale of a child's hiding place, the wall encourages adventures in the labyrinthine corridors of imaginations. The cocoons endeavor to achieve the implied metamorphosis, by providing sanctuary for imaginative thought, where one may lose oneself in daydreams, or in some great book.

Location: Manhattan, NY, USA
Area: 396m²
Designer: Markus Dochantschi, Erin Sonntag
Design company: studioMDA
Photographer: Roland Halbe

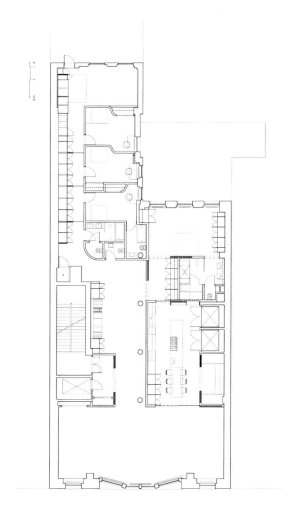

OPTION 4
living/kitchen/dining
bathroom
gallery
guest
nanny/laundry
kids' rooms
master bedroom/bath

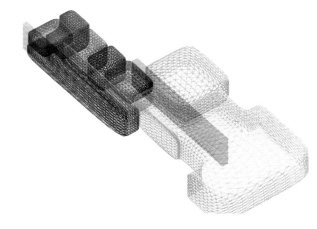

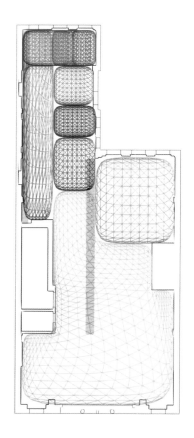

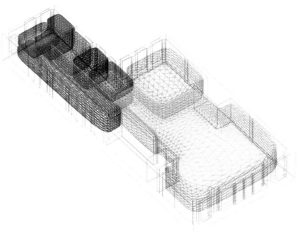

该项目位于曼哈顿诺赫街区的对面，该地区的设计通常采用开放式布局。本案设计是要在客厅区域打造一个廊式景观，将卧室区域分离开来，从而保证各自空间的功能特征与私密性。

儿童房的墙壁给孩子提供了一个独立的空间，里面可以存放玩具，游戏区铺设了Astroturf人造草皮，一张犹如"蚕茧"的双层床铺为休闲和阅读提供了舒适的去处。在富于想象空间的迷宫式走廊里，墙壁采用了大胆的变形设计，一个"蚕茧"式的空间刚好可以容下一个孩子藏身。他们可以在此尽情地幻想，亦或是沉迷于书籍的海洋。

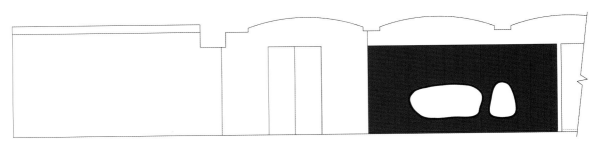

HALL WAY WALL

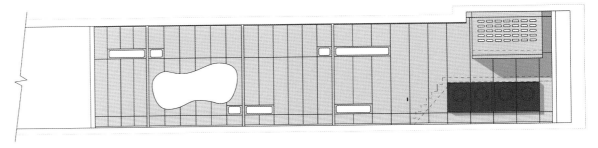

KID'S WALL

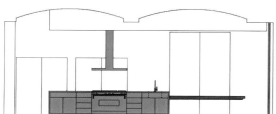

ISLAND - SOUTH ELEVATION

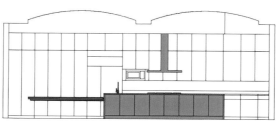

ISLAND - NORTH ELEVATION

NORTH ELEVATION

KITCHEN ELEVATIONS

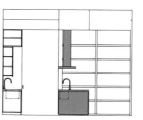

EAST ELEVATION

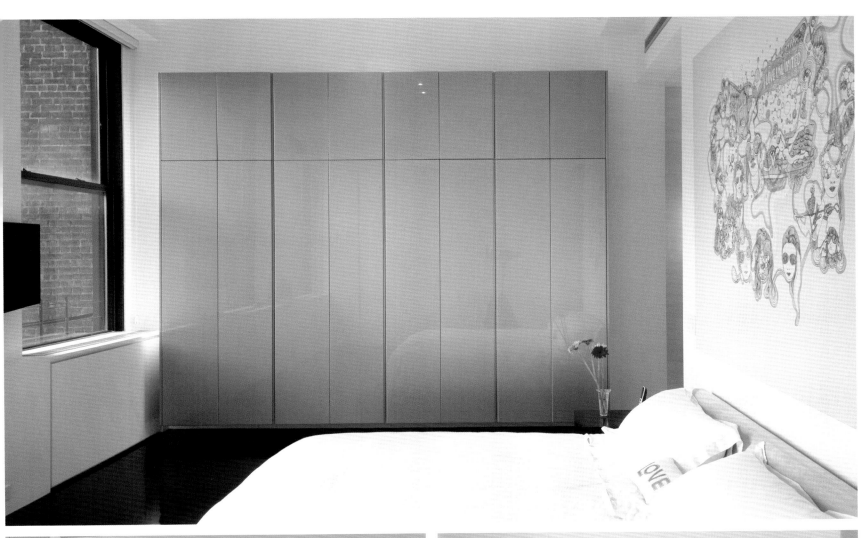

HOME DÉCOR
Ideas for Interior Space, Function & Color

Georgian Townhouse
乔治亚风格的联排别墅

The house is grade II listed so there could be no removal of the original features. No significant work was done to the structure of the house, therefore, other than to knock through a wall between the master bedroom and master bathroom to free up space and to add a dressing room. However, from a decorative point of view, everything else was redone. 'To be honest, the existing space was disgusting,' says Crosland in no uncertain terms. 'It had been untouched for a good 20 years and was very dark. There were ugly cupboards everywhere that needed ripping out and there was even a dumb waiter.'

An early design decision in the hallway provided an inspirational starting point. The designer chose Cactus Paisley paper, designed by Neisha Crosland, and its taupe, brown, gold and silver tones inspired the muted palette for the whole house. The designer wanted to ensure the interior was warm and cozy, so the designer used thick plush carpets wherever possible as well as hotter colors in smaller spaces, such as the greens in the guest bedroom, which also has a red headboard.

Location: London, UK
Designer: Charlotte Crosland

这是一栋二级保护建筑，原始特征不宜改动过大。设计师没有对房屋的结构做大的改动，只是打通了主卧和主卧浴室之间的墙面来扩展空间，增加出一个衣帽间。然而，从装饰的角度看，许多地方都已经改头换面。设计师克罗斯兰直截了当地说："坦率讲，目前的空间很令人厌烦。设计已经落后20年，这里光线暗淡，到处都是一些难看的柜子，需要拆掉，这里居然还有一个送餐旋转架。"

玄关的早期设计方案为整个设计提供了灵感。设计师选用Neisha Crosland设计的仙人掌涡纹图案的墙纸，其上的灰褐色、褐色、金色和银色激发了整个房间的柔和色调。为了确保室内空间的舒适与温馨，设计师尽量使用厚重豪华的地毯，并且在较小的空间内使用暖色调，比如，客卧使用了绿色的地毯，搭配红色的床头。

HOME DÉCOR
Ideas for Interior Space, Function & Color

Concrete House
混凝土住宅

The first sensation that this house produces when people go into the plot is that the building seems to be hidden between concrete walls and vegetable ramps extend up to the roof. They are dyed in dark gray and contain, between them, vegetation areas that seem to climb towards the sky. The house's faade show a spectacular organic view of the whole house and so even the hard concrete shows its most kind face.

The back front of the house is totally opened towards the garden where the lounge, dining room, library, study and bedrooms are. In this facade the wide windows, the volumes set and the projections (made of concrete too), these elements cover the several house's porches. The large window of the main lounge hides itself automatically in order to make this stay completely opened to the exterior areas. The plot includes also an elegant garden, a small lake and a pádel track.

The ecological aspect is very in this A-cero's work: the facade and the roof have the main ecological roles because they are covered with low consume vegetation. Furthermore, the house's roof has been implemented a renewable energy system made of wide surfaces with solar tubular collectors to allow the energy autonomy of the house.

Location: Madrid, Spain
Area: 1600m²
Design company: A-cero Joaquin Torres architects
Building company: Helio Construcciones
Photographer: Luis H. Segovia

　　步入该区域，这栋单户住宅给人的第一感觉便是"隐匿于一片绿色坡地和混凝土墙"中的豪宅。深灰色的外墙，之间是延伸至屋顶的绿地，仿佛延至天际。住宅的外观宏伟大气，给人以强烈的视觉冲击，即便是坚硬的混凝土也展现出了最柔美的一面。

　　房子的后侧正对花园，那里有休闲室、餐厅、图书室、书房及卧室。正面的落地大窗、混凝土块状结构，以及悬挑出的部分，这些元素构成了建筑的门廊。主休闲区的隐形大窗让这里完全开放，外面的景致可以一览无余。整片区域还包括一个优雅别致的花园、一个小湖泊和一条pádel小径。

　　环保是这个设计的亮点，建筑的正面及屋顶覆盖了低耗能的绿色植被，起到了主要的环保功效。此外，屋顶还应用了由太阳能收集管制成的可再生能源系统，为房屋提供能源供给。

HOME DÉCOR
Ideas for Interior Space, Function & Color

Winnington Close
惠宁顿洋房

Entrance
One step in and you'll be amazed. It's not just the glamorous style of the ebony stained oak floor, Bagatelle nickel door handles and Megaron Dressoir Console in white. Cascading from the very top of the building is a stunning crystal chandelier. The circular shape and intricate beading is exquisite.

Dining room
The dining room centre piece is the Sea Break in the Clouds dining table from Paul Glover. The unusual silver finish to the accompanying Asiatides chairs, the grand white chandeliers and large Sea Mirror from India Mahdavi creates a lavish and luxurious scene. Handmade timber design folding doors open into the living room.

Living room
This treasure trove of design delights is practical as it is beautiful. There's a custom designed fireplace from B&D Design and above, within a striking black mirror screen, is the HD-ready 46 inch TV. The Monpas side tables, impressive Paul Glover circular-tiered coffee table with chrome inlaid bowl contrast perfectly with the soft fabrics of the two large sofas, Assago custom-coloured carpet.

Master bedroom
The calming neutral colour scheme of the house continues into the Master bedroom. Overlooking the garden, each furniture piece in this light-filled room is divine. From the kingsize platform bed and matching bedside tables to the Credenza Console and Christopher Guy stools with silver wooden legs, there's no compromise in style. Underfoot is the luxurious and beautiful handmade carpet in Banana silk by Rimo. The lighting is particularly striking and includes Vaughan rock table lamps with Carlisle ivory linen shades and on the walls, Kelly Hoppen IPE Cavalli Amanda lights finished in crystal.

Designer: Kelly Hoppen
Design company: Kelly Hoppen Interiors
Photographer: Mel Yates

门厅
步入门厅，你便会惊叹不已。黑檀色的橡木地板，Bagatelle的镍合金门把手以及白色的Megaron Dressoir门厅装饰台，彰显的不只是迷人的风格。一盏绚丽的枝形水晶吊灯悬于门厅顶部，错综复杂的水晶珠球玲珑剔透。

餐厅
餐厅的中央是Paul Glover设计的Sea Break in the Clouds餐桌。配套的Asiatides餐椅采用与众不同的镀银漆工艺，连同白色的枝形吊灯，以及出自印度设计师Mahdavi之手的大面Sea Mirror镜子，一并呈现出一番奢华景象。纯手工打造的木质折叠门通向客厅。

客厅
客厅融入的奢华设计元素既实用又美观。特别定制的B&D Design壁炉，上方是一块醒目的黑色镜面，内嵌一台46英寸的高清电视机。Monpas边桌，圆形的Paul Glover叠层茶几，再配上铬合金的装饰碗，给人留下深刻印象。这些与两边柔软的布艺大沙发以及Assago定制颜色的地毯形成完美的对照。

主卧
宁静的中性色设计延伸至主卧。从这里可以俯瞰花园，房间光线充沛，每件家具都可谓完美无瑕。从超大的平板床及配套的床头柜，到储物柜以及带有银色木腿的Christopher Guy圆凳，丝毫不失其独有气派。地面铺设了Rimo手工制作的Banana真丝地毯，奢华而又高雅。照明设施尤其引人注目，包括配有卡莱尔象牙色亚麻布灯罩的Vaughan可摇摆式床头灯，以及水晶粉涂层的Kelly Hoppen IPE Cavalli Amanda壁灯。

HOME DÉCOR
Ideas for Interior Space, Function & Color

Kenig Residence
肯尼格住宅

The program of spaces was organized by floor. The basement was finished to become a casual family den-type room with a pool table and couches, etc. The ground floor level was opened up to become a loft-like living room and kitchen with an exterior wood deck floating outside, over the backyard.

The second floor became the girls' floor with bedrooms, a study, kid-lounge and bath. Bachelor retreat was the idea of the third and top floor, with a home office, bedroom and bath.

Tying the home together, the stair acts as a spine from basement to third floor. We were interested in allowing this to read as a distinct element and so structured it to rise free of the adjacent walls. A new skylight was inserted into the roof, above the stair, accentuating the vertical rise and bringing natural light down into the home. The wall behind the stair was finished with sheets of blackened steel that run continuously from the parlor level to the roof. Display elements can be easily affixed to the wall using magnets. By creating an entirely flexible "canvas of metal" we offered a space with infinite possibilities for visuals. The flexibility of the magnet wall fosters the creation of zones organically, as people affix items of interest to them at their own floor - alternately the entire stair could be rigorously created. We felt that the flexibility of the display wall mechanism was appropriate particularly for a family so rooted in a retail tradition.

Location: Brooklyn, New York, USA
Area: 293m²
Designer: James Slade, Hayes Slade
Design company: Slade Architecture
Photographer: Jordi Miralles

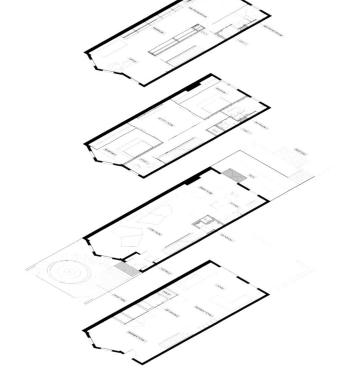

整栋建筑的内部空间区域划分合理有序。最底层的地下室是一间地道道的"密室",摆放了一张台球桌和几张沙发。一楼的双层开放空间是客厅和厨房,厨房的地板延伸至室外,形成一小截露台,与屋后的庭院相连。

二楼是女儿们的卧室、书房、儿童休闲区和浴室。三楼(顶楼)是属于男主人的私密区域,包括办公室、卧室和浴室。

楼梯像脊柱一样将建筑的上下四层串联起来,从地下室直至三楼。这样的结构使楼梯独立于邻近的墙面,可视其为该设计中别具一格的元素。楼梯的顶部新增了一个天窗,拉伸了垂直高度,将自然光引入屋内。楼梯背后是一整片包裹了黑色钢材的墙面,随楼梯一同贯穿客厅至顶楼,其上用磁铁吸附着诸多小物件。通过营造一个多变的"金属画布",向人们展示了一个无限创意的视觉空间。当人们将自己中意的物件贴在相应楼层的墙面上时,整面墙的灵活性便会促成一个个创意区域的呈现。我们能够感觉到,这面展示墙恰好独特地展现了主人从事零售业的家庭传统。

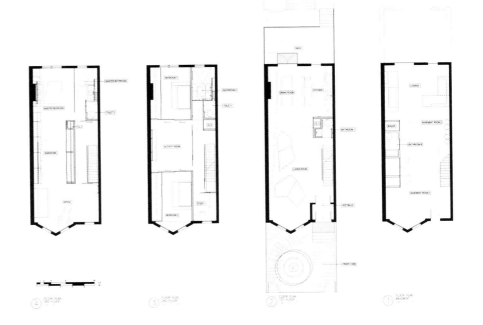

HOME DÉCOR
Ideas for Interior Space, Function & Color

Cetatuia Loft
塞塔图维亚阁楼

The first loft level is an open space comprising the kitchen, dining room and living room, the annexes, technical spaces, office and access hall, connected to the second and third levels by a wooden staircase. The daytime area of the loft (located on the first level of the loft) is organized on two parallel lines, kitchen-dining room, united through the design of the ceiling, and living room-circulation areas. The differences in height and the arrangement of furniture emphasize the virtual separation between the two regions. Windows reach to the floor, which makes the room appear larger, flooded with light, while roof windows bring daylight deep down into the living room through the glass flooring. The stair was reduced to its minimum-the steps. Railing was almost dematerialized by transparent glass, thus non-essential elements that altered the clear perception of the space where eliminated as much as possible.

The glass wall on the last floor for the master's bedroom opens a new direction to the living room and farther to the city through the roof windows. Without these glazed surfaces-horizontal and vertical-the house would have lost a very important spatial dimension and the relationship between the interior and the landscape.

Strongly emphasized verticality of the living room walls is attenuated by the gallery of openings of various sizes and depths, connecting this room to something beyond the wall, either inside or outside the house. These are actually the interior decorations of the house.

On the second floor there is an extension of the living room with fire place, sleeping room, bathroom and storage. The last floor is for master's bedroom and a generous bathroom opened to the outside through a series of roof windows and a partially glazed wall.

Location: Brasov, Romania
Area: 415m²
Designer: Ion Popusoi, Bogdan Preda
Design company: In Situ
Engineering: Rozini
Photographer: Cosmin Dragomir

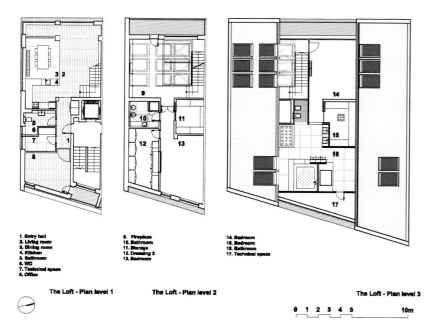

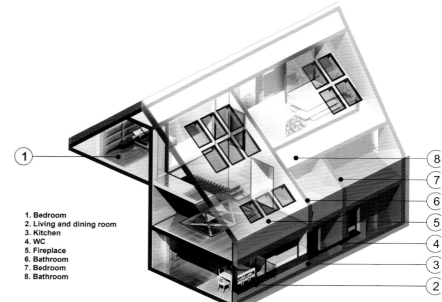

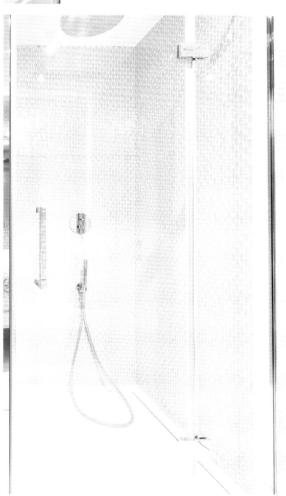

阁楼第一层是一个开放的空间，包括厨房、餐厅和客厅、附属建筑、设备区、办公室和入口大厅，由一个木制楼梯连接到二楼和三楼。

日常活动的阁楼区域（第一层的阁楼空间）被分成平行的两部分：厨房和餐厅直通天花板，与客厅区域相通。空间高度的差异与家具的布置放大两个区域之间的虚拟空间。

落地窗使房间显得更加宽敞，室内阳光明媚，屋顶的天窗也使阳光透过玻璃地板直抵室内深处。楼梯被简化为几级台阶。透明玻璃几乎取代了楼梯扶手，从而尽量消除了那些改变人们空间感知的不必要因素。

最后一层的主卧的玻璃墙直通客厅，透过屋顶天窗可远眺整座城市。如果没有这些或水平或垂直的玻璃表面，房子就会失去非常重要的空间感以及室内外景观的融合感。

无论是室内还是室外，阁楼的层高和空间大小弱化了对客厅垂直墙壁的过度强调，将这一区域与其他区域融为一体，这实际上是内部装饰的效果。

阁楼第二层是客厅延伸的部分，包括壁炉、卧室、卫生间和储藏室。最后一层是主卧和宽敞的浴室，屋顶天窗和部分玻璃外墙使浴室朝向室外的开放区域。

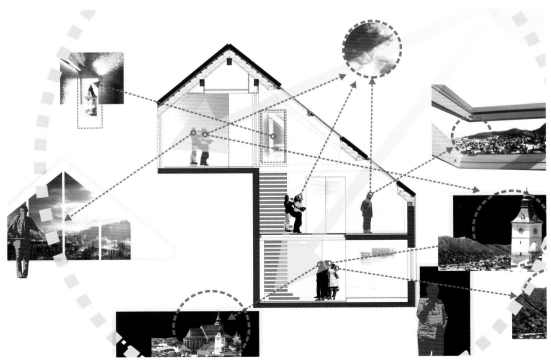

HOME DÉCOR
Ideas for Interior Space, Function & Color

Collins Ave, Miami Beach
迈阿密海滩的科林斯大街公寓

The goal of this architectural renovation and interior design project on Collins Avenue in Miami was to give the clients an apartment with a modern loft feeling, and to open the living spaces to the Atlantic Ocean views. The clients wanted to see ocean views from all the rooms from this Art Deco apartment house. The subtle and minimal architectural envelop of white walls and limestone floors became the perfect backdrop for the client's modern art collection, and unconventional mixture of furniture. Prominently placed in the foyer is a large red sculpture by Verner Panton, the actual artist's prototype, setting the eccentric quality of the collection of furniture and artwork. Modern custom design furniture by D'Aquino Monaco and classical pieces are irreverently mixed. The crystal dining table floats and disappears reflecting the ever-changing light. Plaster walls are softened by curtains fabricated out of parachute cloth, with down lighting softening the perimeter of all rooms.

Structural columns and beans were exposed, plastered, and in some cases gold leafed within this neutral setting. Artful triangular cut outs expose structural beams in the ceiling and become overhead lighting fixtures as well, as is visible above the dining table.

While the main living area is an open floor plan, the master and guest bedroom are secluded from the rest of the apartment. An oversized plaster and gold leaf headboard designed by D'Aquino Monaco serpentines through the bedroom somewhat reminiscent of a Dali painting, our homage to the exuberance of old Miami Beach.

Location: Miami, USA
Designer: Carl D'Aquino, Francine Monaco
Design company: D'Aquino Monaco Inc.
Photographer: Marnie Steele

　　这个建筑翻新兼室内设计项目位于迈阿密的科林斯大街，设计的目的是要向客户呈现一个现代化的阁楼风格的海景房，生活空间面向大海，客户要求在这个艺术装饰的公寓的每个房间都能够观赏到大西洋海景。现代气息的艺术品，别具一格的混搭家具，精巧细致的白色墙壁和石灰岩地面，这些元素完美结合。大厅的显著位置摆放了一个由Verner Panton原创设计的大型红色雕塑，突显了家具与艺术品的与众不同。由D´Aquino Monaco定制设计的现代家具与一些古典风格的配件浑然天成。水晶餐桌在光线的映衬下色彩斑斓，千变万化。在灯光下，石膏墙面搭配伞布制成的窗帘显得格外柔和温馨，弱化了空间的界限。

　　设计没有将建筑的立柱和横梁加以隐藏，只是刷上了灰浆，一些地方在整个中性色的背景下点缀了金色。天花板上，人工开凿的三角形开槽暴露出横梁的位置，餐桌上方的一个可用来固定吊灯。

　　主要的起居空间呈开放式设计，主卧和客卧与其他空间分离。由D´Aquino Monaco设计的一个超大的金色石膏床头板贯穿整个卧室，令人不禁想起达利的油画，真实地展现了迈阿密海滩的复古风情。

HOME DÉCOR
Ideas for Interior Space, Function & Color

Vivienda 4
维维埃达4号住宅

The initial idea of this project is to integrate the house in the natural environment of the residential area in which it is. In that way, the house looks to the fantastic views of lakes and green extensions that spread over the common areas of the estate. The structure's house is based on game of two rectangular volumes. They are solving with clean finishes, good materials and exclusive qualities.

The verandas and the pergolas, elements as volumetric as the housing, give personality to the facade. The covering as such disappears turning into a flat roof as expression of the evolution of the creative process. The forms purity moves to the constructive scheme, to the materials and to the environment impelling the union between the building and the context where it is located.

Inside the following stays are located: foyer, room of kitchen, dining room, office, larder, a cleaning of courteousness, seven bedrooms (with his respective baths, two dressing-rooms and a patio). In the later part, across big large windows, there is lounge that is opened for the garden, the swimming pool and for the natural scenery that offers the urban development. It surrounds itself with a landscaping realized with attractive interventions composed by lakes, birds and a big arboreal extension. In the exterior the pedestrian ways and the central landscaped area incorporate a vegetation of indigenous character, resistant to the climate of the area. The stockades of every plot and the trees and shrubs planted along the same ones, they delimit perfectly the transit between the private ownership and the community one.

Location: Madrid, Spain
Area: 800m²
Design company: A-cero Joaquin Torres architects

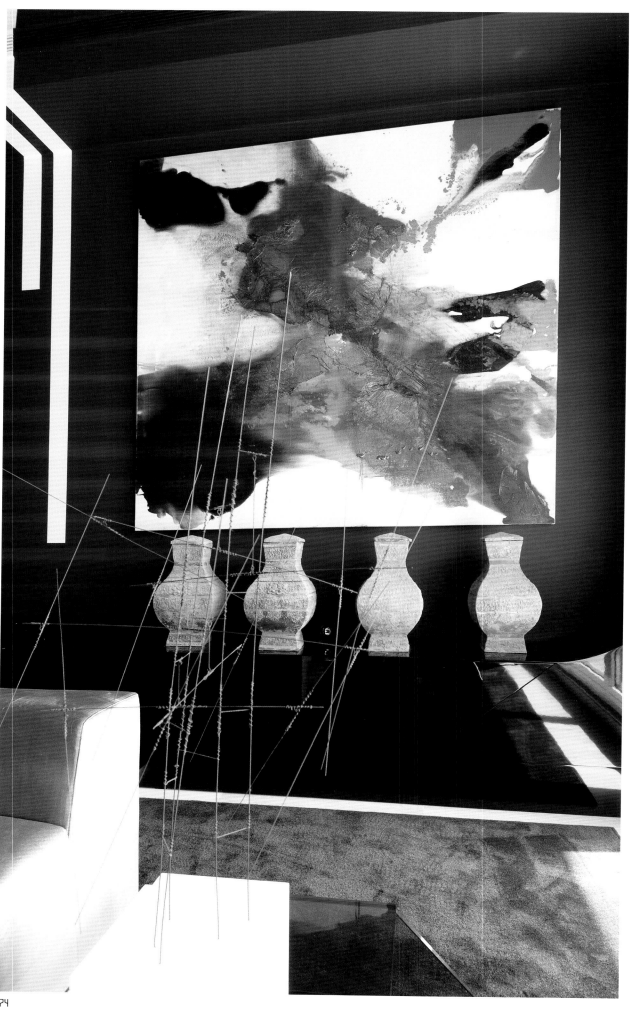

该项目的初衷是想将房屋与住宅区的自然环境融为一体。这样一来，奇妙的湖光景致与绵延的公共绿地风光便可尽览于房中。房子的结构基于两个矩形方块，其表面精整，选材上乘，品质独特。

走廊、藤架以及其他一些方块元素都与房子的造型如出一辙，这些赋予了建筑个性化的外观。草坪与屋顶平位浑然一体，这一设计体现了整个创意过程的演变。这种纯粹的形式转移到了建筑的主题、材料以及与之和谐的自然环境中。

再来看内部设计，包括客厅、厨房、餐厅、工作室、储藏间、七间卧室（带有独立的浴室）、两个衣帽间和一个天井。在靠里的位置有一间面朝花园和泳池的休闲室，透过宽敞的落地窗，可以欣赏到郊外的自然风光。整栋建筑身于优美的景致之中，湖水、飞鸟，以及高大的树木点缀其中。户外的人行步道和中央景区选取了具有本地特色的植被，以抵御这里的气候变化。每个区域的栅栏，以及沿着栅栏种植的树木与矮丛在私有空间和公共区域之间形成完美的过渡。

HOME DÉCOR
Ideas for Interior Space, Function & Color

The Orange Grove
新加坡中区橘林

Design Goal
The design goal was to create an interior that fulfils the pampered lifestyle of the sophisticated, ultra-wealthy, international jet-set market. The designer identified actress/celebrity Michelle Yeoh of Crouching Tiger Hidden Dragon fame, one of Asia's most recognized international celebrities as the "face of the development" as she embodies the best of the East and West lifestyle.

Design Approach
The concept was to treat the apartment as a luxurious modern tropical Asian bungalow in an urban setting. The designer created a strong sense of place by referencing Asian luxury brands such as Aman resorts and Banyan Tree for the jet-set patriarch/matriarch market.
The interior is both modern and classic, taking influences from the East and the West. The designer imagined the occupant to be a collector of both traditional Chinese blue and white porcelain and of modern contemporary Asian art. Even the shimmering pink walls of the family room take its inspiration from famille rose porcelain and the Chinese cabinets in the living room are wrapped in stainless steel.

Location: Orange Grove Road, Singapore
Designer: Cameron Woo
Design company: Cameron Woo Design

设计目标

营造一个富丽堂皇、精致奢华的内部空间,满足国际化高端客户的需求。设计师邀请亚洲最具国际影响力的演员,电影《卧虎藏龙》的女主角——杨紫琼作为该项目的形象代言人,她是将东西方极致生活方式完美结合的典范。

设计方法

设计理念是要在市区打造一个现代奢华且富有亚洲热带风情的别墅式公寓。设计师还参考了亚洲专门面向高端客户的豪华酒店品牌——安曼酒店集团和悦榕庄酒店集团的设计来打造浓烈的地域感。

其内部设计融合了东西方的古典与现代。设计师设想,房子未来的主人会是一个收藏爱好者,藏品里既有中国传统的青花瓷,又有现代的富有亚洲特色的艺术品。卧室粉色亮光墙面的设计灵感就源于粉彩瓷器,客厅里的中式博古架则使用了不锈钢表面。

HOME DÉCOR
Ideas for Interior Space, Function & Color

Sentosa House
圣淘沙岛住宅

The house was situated on a corner lot with two sides facing a water canal. The architectural intent to celebrate this unique waterfront view was achieved by strategically locating a free-standing oval shaped living room that anchored the project on the site, orientating it towards the water. The rest of the house is contained within a fluid and natural form and serves as a backdrop to living room.

Every architectural decision was focused around the site and the views it provided. With half of the project open to the canal, the waterfront became a focal point around which the house was designed. Each room in the house is orientated to face the water, and both fully exploits and celebrates the views of the waterscape.

The aim of interior design was to work with the architecture, emphasizing the courtyard and ensuring harmony between the interior and exterior. It was important that the two melded with each other, creating a seamless flow of circulation that did not simply occur laterally from outdoor to indoor, but vertically between floors. In this manner, the entire house becomes an endless playground that once meanders through; always open to views of the water, yet intensely private at the same time.

Location: Singapore
Area: 520m²
Designer: Rob Wagemans, Lisa Hassanzadeh, Melanie Knüwer
Design company: concrete architectural associates
Photographer: Sash Alexander, concrete architectural associates

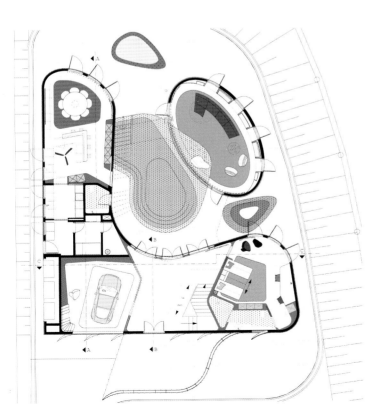
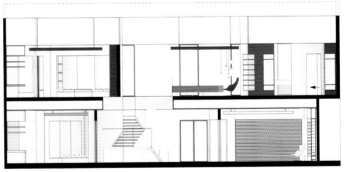
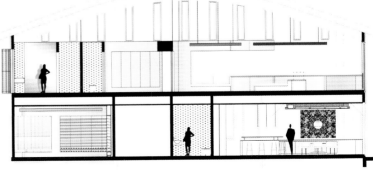

section AA

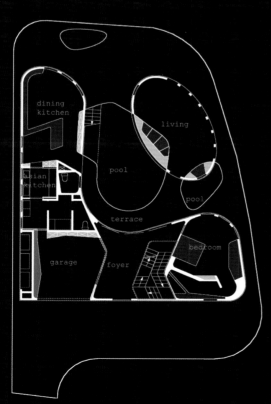

本案位于河道转弯处,两面临水。建筑学理念是通过一个椭圆形的直接朝向运河的独立起居室,来突出建筑周边水畔优美的风景。建筑中其他房间随流线型自然分布,作为背景衬托起居室的独特地位。

每个建筑的细节均围绕观赏风景的位置和角度而精心设计。整栋建筑有一半直接面向运河,河畔区域是建筑设计的核心。建筑中的每个房间都要直接面向河畔,最大限度地充分利用如画般的景致。

内部设计的主要目的是同建筑一起突出庭院,确保内外空间的和谐一致。如此设计装饰最重要的环节就是如何将内外空间有机融合,不仅仅是简单地在平面上从内到外,还要在垂直空间上营造出浑然一体的效果。如此一来,整个居所成了一个无尽的操场,漫步其中,满眼尽是水畔风光,却也极具私密性。

HOME DÉCOR
Ideas for Interior Space, Function & Color

Isabel
伊莎贝尔

The residents, a couple and their granddaughter, approved all the proposals made by the architect that, not letting aside classic style, she harmonized colors, modernity and incorporated playful to the project, that counts with Campana brother's work, using a lot of glass and crystal. Functional, all the apartment is automated by Cynthron. The audacity in the place is noticed since the entrance, with a hall all covered by glass – executed by Penha Vidros – that makes the space feels bigger than it is. In the kitchen the option was to make a creative and innovative ambient that toys with the playful. The walls received coating with pink glass, from Penha Vidros. The room received paving from Tampos thassos, Gyotoku, and cabinets from Florense. The place has modern equipments from Viking Eletrodomésticos and also from Iesa.

The office counts with the color element, predominating red. The tissue of the curtain is from Empório Beraldin, executed by Arte Markante. In the same tone, a couch from Firma Casa.

The living is integrated, by running doors, to the balcony, dinning room and home theater. For this reason, a sober decoration was made, predominating light colors that matches with the other rooms. Lafer's reclining armchairs, coating by artificial fur made by Celina Dias, were placed in front of a big screen, guaranteed comfort to the viewers. The center table was all made in massive acrylic. The couch is from Montenapoleone. A mirrored panel, besides having the screen, guarantees a clean room, because it is also used to hide wires.

The same refinement is founded in the master room. The bed and the armchair, both in classic style, are from Montenapoleone. The wall is highlighted by the wallpaper from Wallpaper, which counts with golden details that matches with the furniture.

Location: São Paulo – Vila Nova Conceição
Area: 500m²
Designer: Brunete Fraccaroli

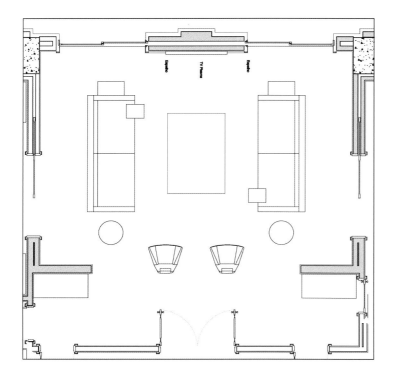

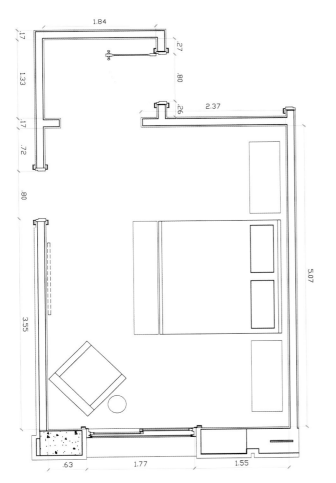

设计师没有舍弃古典的设计风格,同时将色彩与现代元素和谐地融入到Campana兄弟的设计项目中,运用了大量的玻璃和水晶。对此,房主夫妇和他们的孙女十分赞赏。从功能方面看,所有房间都通过Cynthron实现了全自动化。大胆的设计尝试从门口开始,门厅使用了Penha Vidros设计的玻璃装饰,这让整个空间显得更加宽敞。厨房的设计试图营造一个富于创意的、新颖的空间,让人感到身心愉悦。墙壁铺装了Penha Vidros的粉色玻璃。房间的地板和橱柜分别出自Tampos thassos、Gyotoku和Florense。此外,还配备了一些Viking Eletrodomésticos和Iesa的现代化设备。

书房也运用了色彩元素,以红色为主色调。出自Empório Beraldin的薄纱窗帘由Arte Markante亲自设计。Firma Casa设计的长沙发椅也选用了同样的色调。

一扇滑动门将客厅、阳台、餐厅和家庭影院有机结合。为了与其他房间搭配,客厅选择了素雅的装饰风格,配以柔和的浅色调。大屏幕前摆放了一把Lafer's扶手椅,配以Celina Dias的人造毛皮装饰,使观看者更加舒适惬意。中间的桌子由大块的亚克力板材制成。沙发的设计出自Montenapoleone之手。一块类似镜面的板材使房间显得干净利落,不仅可以当做屏幕使用,而且还可以遮挡电线。

主卧的设计同样雅致精巧。古典风格的床和扶手椅均由Montenapoleone设计。墙上的Wallpaper壁纸是房间的一大亮点,金色的细部设计与家具完美搭配。

HOME DÉCOR
Ideas for Interior Space, Function & Color

Riverside Drive
河畔大道公寓

To fulfill the client's desire to highlight views of the Hudson, the designers created a kinetic design of sliding lacquered wall panels in this Riverside Drive apartment of approximately 223 square-meters. The clients came to D'Aquino Monaco Inc with a clear desire for a palette that was black and white, with a little gray thrown in. They were ready to make a big change in their lives and to this space.

Every detail was chosen by design in a deliberate attempt to create a fluid, reflective space. The river view now provides the color within this apartment while varying textures of plaster, lacquer, and matte surfaces create depth and interest within the almost all-white palette.

A central corridor was removed to allow the living areas to flow into one another while drawing the light and views of the Hudson River further into the space. Spaces are defined by carved polylaminate sliding doors and dropped ceilings. The flexibility of the living spaces combined with rich textures and a play between matte and polished surfaces allows for a truly dynamic space and a fresh setting for contemporary art and furniture.

Location: New York, USA
Area: 223m²
Designer: Carl D'Aquino, Francine Monaco
Design company: D'Aquino Monaco Inc.
Photographer: Peter Margonelli

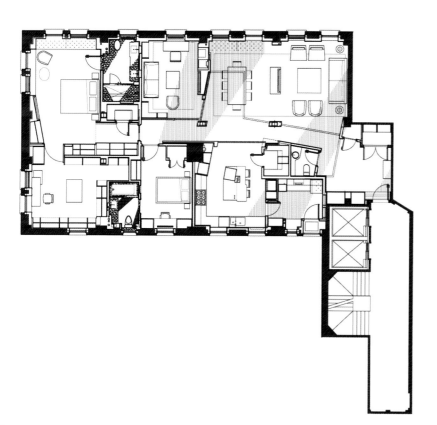

该项目总面积约为223平方米,为了满足客户欣赏哈得孙河风景的要求,设计师采用了富有活力的移动式喷涂墙面板材。客户找到D´Aquino Monaco设计公司,明确了以黑白为主色调,加以些许灰色点缀的设计风格。他们准备对这个空间和他们的生活进行大的改变。

设计师精心打造每一处细节,试图营造一个流畅、富于遐想的空间。河畔美景为设计平添了几分姿色,石膏、漆面、哑光表面,这些不同元素让这个几近完全白色的空间更具层次感和趣味性。

将原来中心区域的走廊拆除,扩大了客厅空间,这样,居住者可以尽情享受自然的阳光以及哈得孙河美景。雕刻的多层移动门和吊顶将不同空间加以区分。客厅的设计富于变化,不同材质相得益彰,哑光与亮光的不同搭配完美结合,这些元素组成了一个真实的动态空间,为现代艺术和家具陈设提供了一个清新的背景。

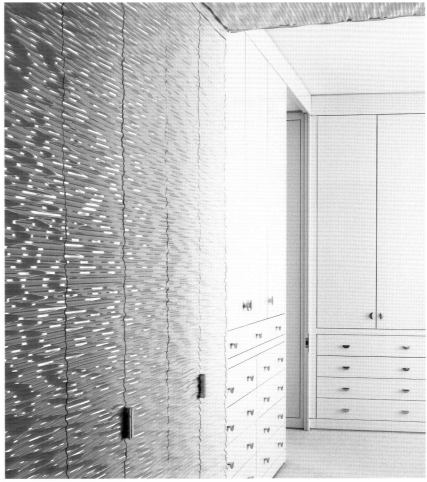

HOME DÉCOR
Ideas for Interior Space, Function & Color

The Lake by Yoo
Yoo湖区

Set around a network of six lakes spread over 263 rural hectares, the Lakes by Yoo is a joint venture with The Raven Group, led by Chairman Anton Bilton and international design, branding and property investment company Yoo. This project is a unique combination of beautiful country living with all the urban luxury you could imagine – from stylish interior design to a concierge service to prepare your home and stock your fridge on arrival.
The Lakes by Yoo is a number of firsts all rolled into one development – it's the first time Yoo has gone country (in The Cotswolds), the first time Jade Jagger for Yoo designs have been available in the UK and the first Yoo properties available to short term visitors and holiday makers.
The Lakes by Yoo project was all about adding a bit of urban style to a country setting. The designer used vintage pieces, loose linen throws and gorgeous textures so that anyone would feel instantly at home in the space. The stunning location and oustanding vision of John Hitchcox and the rest of the Yoo team made this a great project to be a part of.

Location: Cotswolds, UK
Designer: Kelly Hoppen
Design company: Kelly Hoppen Interiors
Photographer: Mel Yates

　　六湖环绕，占地263公顷的Yoo湖区是Raven集团旗下的一个合资项目。该集团由安东·比尔顿和国际设计、品牌、地产投资公司Yoo共同领导。该项目是一种优美的乡村生活与超乎想象的都市繁华的独特结合——从独具现代风格的室内设计到接待服务，从住宅安置再到及时的配送服务，可谓一应俱全。

　　Yoo湖区集多个"第一"于一身。这是Yoo第一次入驻乡村（科茨沃尔德），杰德·贾格尔设计的Yoo首次在英国亮相，同时，也是第一次可以让短期游客和度假的人尽享Yoo风貌。

　　Yoo湖区项目是在乡村风格中融入了些许城市气息。设计者使用古香古色的器具、宽松的亚麻家纺和雍容华贵的织物，使人有宾至如归的感觉。绝佳的地理位置和约翰·赫区考克靓丽的风景以及Yoo的其他设施使其成为一个了不起的工程项目。

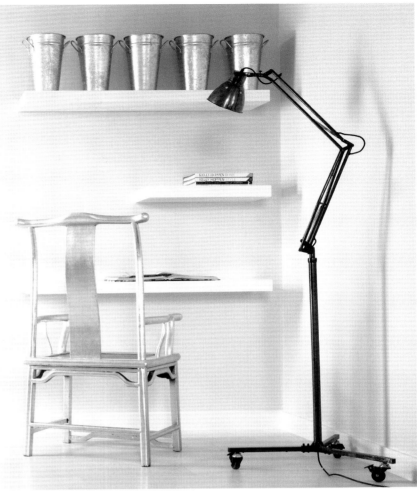

HOME DÉCOR
Ideas for Interior Space, Function & Color

Fractal Pad
叠拼公寓

This project is about a highly glamorized domestic typology: the urban bachelor pad. The client, a very young and successful Wall Street commodities trader wanted to create an entirely interior landscape away from the city and the outside world. The designer's response was to consider the experience of this primary residence as a daily return to Plato's Cave; as the conscious exfoliation of exterior stresses through the creation of an enveloping abstract domestic landscape, where light and shadow could simply be appreciated for their beauty and mystery, and the outside world could cease to exist. The resulting completed project, Fractal Pad, is a sumptuous geometric oasis for a lover of mathematics and geometry. It develops from the formal logic of fractal geometry, bending space and light to create a seamless, harmonious experience. The crisp new interior installation contrasts starkly and elegantly against the loft's original rough-hewn concrete columns and slab, and a rich, blood-red Santos mahogany plank floor, creating a desired effect that is at once akin to a futuristic private luxury aircraft and a primordial cave.

Location: New York, USA
Area: 279m²
Designer: Matthew Bremer, AIA
Design company: Architecture In Formation
Project manager: Paulo Flores
Photographer: Tom Powel

该项目旨在为城市的单身贵族打造一处富于本国魅力的栖身之所。房主是华尔街一名成功的年轻商品交易员，他希望创造一种远离城市喧嚣、与世隔绝的室内景观设计。基于公寓的主要功用，设计师打算将其设计成为一个朝离暮返的柏拉图洞穴：通过有意识地打造抽象的室内设计来与外界压力隔绝，光线和阴影的完美搭配可以直白地展现装饰的美丽与神秘，使外界的喧嚣不复存在。基于上述理念完成的这一作品为喜爱数学和几何学的人打造了一处奢华的几何绿洲。该设计源于分形几何的形式逻辑，利用空间和光线的弯曲创造出一种无缝、和谐的体验。简洁而新颖的室内装饰，原始粗糙的混凝土立柱和楼板，富丽堂皇的桑托斯红木地板，这些元素形成优雅、鲜明的对比。这一理想的设计效果既像一架未来的豪华私人飞机，又像一个原始的部落洞穴。

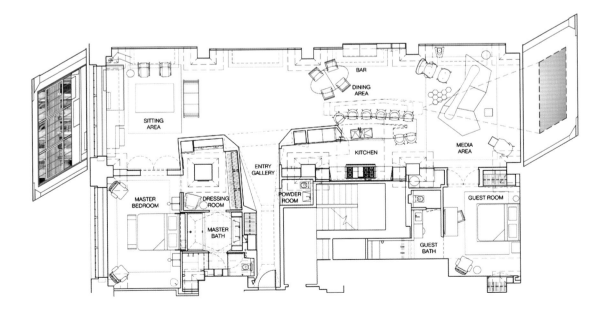

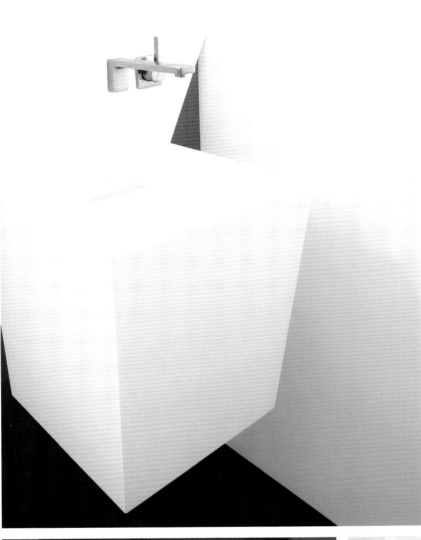
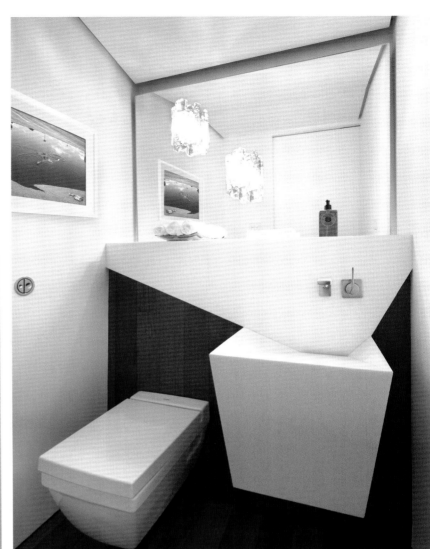

HOME DÉCOR
Ideas for Interior Space, Function & Color

Pied a Terre
临时住所

Helene Benhamou's London pied-a-terre is an homage to her style. Located in Chelsea just behind King's Road, the spacious 250m² traditional English style maisonette is bathed in light and elegance throughout. White walls and parquet floors are the canvas where neutral colours and bold accents come alive.

The heart of the apartment is an alluring, open living space where the sitting room, dining area and kitchen come together in refined harmony. In the sitting room black, gold, soft gray, and taupe are the palette of warm and chic. The straight lines of classic furniture are accented by the slim and delicate curves of fine glass vases and a daring bronze sculpture while paintings with colorful, cubist-inspired forms play on the walls. Elegant molding puts the finishing touch on this clean yet sumptuous space.

The adjacent dining area mixes the formal and the relaxed. A wall lamp with round, gold bubbles presides over a low and circular dark wooden table on which a silver and ivory tea service rest, inviting you to sit down and stay for a cup. Helene brings the tranquil tone and colours of the sitting room to her luxurious bedroom. The air here is of retro sophistication. It is a place of purity and repose where marble, white walls and sleek silver fixtures capture the essence of clarity and peace.

Location: London, UK
Area: 250m²
Designer: Helene Benhamou
Design company: HB Interior

海林·本哈默在伦敦的临时寓所忠实地再现了她本人的风格。该寓所位于切尔西、国王路的正后方，面积为250平方米，是一栋传统的英伦风格公寓，明亮通透，高贵典雅。白色的墙壁和木质地板都搭配了帆布装饰物，中性色调配以黑色边框，栩栩如生。

公寓的中心是一个迷人的开放式起居空间，其中的客厅、餐厅和厨房浑然一体，尽显精致与和谐。客厅运用了黑色、金色、浅灰色和灰褐色这些温暖而别致的色调。古典家具的笔直线条，精美玻璃装饰瓶纤细柔美的曲线，大胆创新的青铜雕刻艺术品，墙上悬挂的立体派彩绘，这些元素相互映衬。雅致的装饰线条给这个简洁而华丽的空间平添了几分精美。

与之相邻的餐厅融合了正式与休闲的风格。圆形的金色泡状壁灯，搭配深色的木质圆形餐桌，桌上摆放着银色和象牙色的茶具，邀您来此休息片刻，小沏一杯。

海林还将起居空间的这份恬静风格与色调带入到她豪华的卧室。这里颇具古典的奢华，是纯粹的养身之所，理石、白墙、光亮的银色摆件，这些元素凝聚了清新与安宁的精髓。

HOME DÉCOR
Ideas for Interior Space, Function & Color

The House Of Cesar, Roberta and Simone Micheli

西萨、罗伯特和西蒙·米歇利的家

The genesis of the plan of the House in Florence by Simone Micheli where he lives together with his wife Roberta and his son Cesar lies its foundations in the concept of modern luxury the Tuscan architect developed after a thorough and targeted thinking. It is a residence made up of 90% eco compatible materials and it is an authentic hyperrealist portrait of the 'Ethical Luxury' which is one of the main focuses of his daily architectural searching. It is a dynamic, extremely fresh and vivid intervention taking place in an ancient 1800 setting which has converted these spaces laden with memories into a new environment capable of hosting meaningful fragments connected with a fast and unstoppable metropolitan life. In this way the full-height space which breaks the residential distribution rules is divided by a big brick arch, it features an unconventional distribution pattern on the ceiling and large openings revealing the opposite garden. It hosts the episodes of daily life in succession while a wedge, a low height diagonal volumetric element hosts services and the kitchen. Above this splinter which can be reached by walking the staircase we find a space for children to play marking the end of the whole visual description pattern. A spiderweb, a thread texture watches over the floor and creates an informal and valuable frame which embellishes the body of the services in an elegant and light way.

Location: Florence, Italy
Area: 200m²
Designer: Simone Micheli
Design company: Simone Micheli Architectural Hero
Photographer: Juergen Eheim

这是设计师西蒙·米歇利和妻子罗伯特及儿子西萨在佛罗伦萨的家。该设计源于这位托斯卡纳建筑大师经过充分并且有针对性的思考，进而形成的现代奢华概念。整套房子90%的建筑材料为生态环保材料，它向人们描绘了一个地道的超现实主义，"符合道德传统的奢华"，这也正是西蒙·米歇利所追求的核心建筑理念之一。这栋19世纪的建筑已经由一个满载记忆的空间变成一个全新的住宅，有意义的记忆碎片与永不停歇的快捷的都市生活和谐统一，融入了动感的、极富活力与创造力的元素。一个大型的砖砌拱门将打破住宅分区规律的全高度空间一分为二，天花板上的这个非传统的分区结构成为设计的一大亮点，通过宽阔的拱门开口可以看见对面的花园。开阔的空间将日常生活的片段串联成一个整体，而一个嵌入式区域和一个对角单元的下方分别被设计为厨房和餐厅。沿着楼梯来到对角单元的顶端，可以看到了一个儿童活动区，这是整个可视化描述模式的终点。通过网状的装饰线条结构可以俯瞰地面，这种不规则的但又颇具视觉效果的网格构架以一种优雅轻盈的方式修饰了餐厅的主体部分。

HOME DÉCOR
Ideas for Interior Space, Function & Color

Walmer Loft
瓦尔默阁楼

Overlooking Toronto's historic Casa Loma Stables, this 1920s duplex was converted from a group of small, dark rooms into a graphic designer's residential loft. The renovation was designed to retain the charm of the building's historic exterior, while transforming its interior for contemporary living with clean lines, open space, increased natural light and specifically framed interior and exterior views.

Most of the interior walls were removed and the living room ceiling was opened up to the roof rafters, creating a double-height loft space that feels much larger than its modest floor area. A mezzanine was inserted to create a private zone that can be accessed by a floating mahogany stair with a light steel railing. Small windows were enlarged to infuse the space with natural light, and a large two-storey window was added to frame a view of the stables and to give the feeling of floating in the tree-tops. Built-in elements such as the fireplace and custom-designed display areas create a sense of division in the otherwise open plan, and frame curate views that allude to the owner's artistic outlook.

The owner's Asian heritage is reflected in the interior's warm red woods, white walls and crisp geometric lines, providing a clear backdrop for the owner to showcase his long time collection of Asian art and artefacts and classic modern furniture.

Area: 153.29m²
Designer: Heather Dubbeldam, Tania Ursomarzo, Alex Lam, Katya Marshal
Design company: Dubbeldam Design Architects
Construction: M+K Construction
Photographer: Tom Arban

历史上著名的多伦多Casa Loma Stables建于20世纪20年代，这栋由许多狭小、黑暗的房间组成的二层小楼如今已经被改造成为一个平面设计师的公寓。在翻新过程中，建筑本身具有历史意义且极富魅力的外观得以保留，而室内设计却是现代风格，清晰的线条，敞开的空间，充足的自然光线，特别是通过玻璃窗连接的内外空间，犹如一幅幅风景画。

建筑的绝大多数内墙被拆除，客厅的天花板延伸至屋顶的椽子，形成双层高的阁楼空间，使房屋面积更加开阔。中间的夹层分离出一块僻静空间，一段配有轻钢扶手的红木楼梯通向这里。扩大了原有的小窗户，目的是改善采光，此外，还增设了一个双层高的大窗，形成这里的一景，窗子与屋外的树梢齐高，让人仿佛浮于其上。嵌入式的陈设，例如，壁炉以及特别定制设计的陈列区域使敞开空间有了区域的划分，同时，这些元素也体现出主人的艺术品位。

室内暖色调的红木家具、白色的墙壁以及清晰的几何线条反映出主人的亚裔文化传统，这些背景也突显了主人长久以来收集的具有亚洲特色的艺术品、手工艺品以及典雅的现代家具。

HOME DÉCOR
Ideas for Interior Space, Function & Color

White Apartment
白色公寓

The design designers settled upon was an intervention where the white color is dominant and plays the role of a clean and immaculate background on which the main theme of the apartment is evolving-the furniture, which is treated as a unitary contemporary insertion. The effect of the interior design is based on the contrasts between support/insertion, old/new, permanent/temporary. The basic themes of the design are the rhythm of the paneling of the furniture pieces, the cuts and the cut-outs within them that follow the rhythm and the modulations of an idea of musicality. The furniture and partition units define areas and organize the space, ordering the whole display of activities. By its design, the furniture overcomes its state of static object within space and takes part in a dynamic manner in the definition of the apartment.

From the former structure of the apartment the designers maintained as a main feature the dynamic longitudinal wall that separated the two living areas (diurnal/nocturnal)-a structural element that integrated niches for depositing and passages. It is "wrapped" in the new furniture and transformed into a functional volume that takes part in the interior definition of space, a contained that plays the role of a space divider.

Location: Romania
Area: 150m²
Designer: Baldea Maja, Wenczel Attila, Toma Claudiu
Design company: PARASITE STUDIO

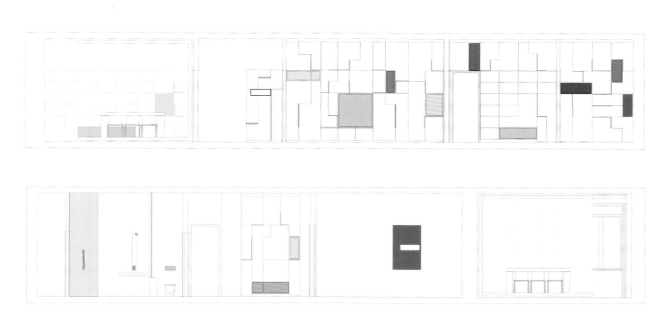

该项目的设计方案最终确定了以白色为主的基调,在干净利落的白色背景下形成了公寓的设计主题——现代的整体嵌入式家具。整个室内设计的效果是基于支撑物与内嵌家具、新与旧、永久与暂时的对比。其主旋律为镶嵌饰板的家具拼接,板面切割所形成的图案演绎出娴熟动听的乐章。

家具和隔断定义、划分了空间,使日常活动合理有序。这种设计令家具摆脱了空间上的静态束缚,使之成为公寓设计的动态元素。

在原有结构的基础上,设计师将动态的纵向隔墙作为主要特色,将白天和晚上两个起居空间分离开来。这一结构还将储物格间与过道整合于一体,用全新的家具包装后形成一个功能模块,起到了对内部空间的定义和分割作用。

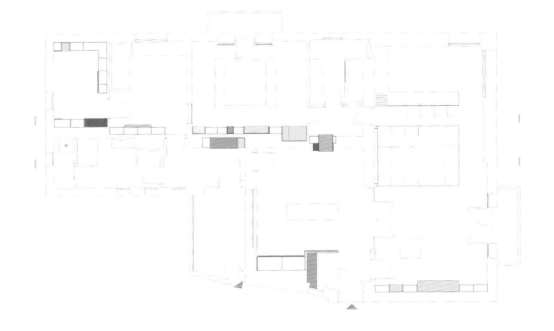

HOME DÉCOR
Ideas for Interior Space, Function & Color

Absolutely Fabulous
美妙绝伦

The owners of this large 1950s house in the English countryside wanted a total re-design of their property – out with the old and in with the new! They contacted interior designer Adrienne Chinn who proposed extensive changes. The old staircase was removed and a new glass and oak open-tread staircase was built and brought over from Germany to bring light into the dark hallway. The spacious open-plan living and dining room is the star of the house. The dividing wall was removed to create the spacious new room, and Adrienne and her associate designer Zhu Cheng designed a feature TV/fireplace wall with a cantilevered travertine ledge. The clients own many pieces of art glass, so Adrienne designed special shelves throughout the house for the clients to display their collections.

The old kitchen was small and dark, so Adrienne suggested putting in folding glass doors to the beautiful garden, and custom-designed oak and glass kitchen units. All the bathrooms were redesigned and feature coloured Corian, stylish sinks, walk-in showers and a huge Jacuzzi bathtub. The stunning red and white master bedroom has a unique display wall/headboard where the designers included niches in place of bedside tables. The couple embraced the design ideas, packed up their things and moved out as the builders moved in. Seven months later it was all done, and it's just absolutely fabulous!

Location: Seattle, USA
Designer: Adrienne Chinn, Zhu Cheng (朱橙)
Design company: Adrienne Chinn Design
Photographer: Richard Gooding

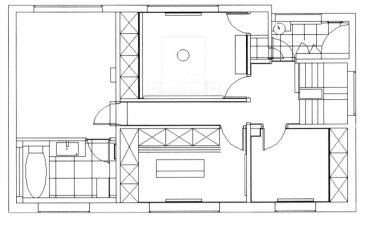

FIRST FLOOR PLAN

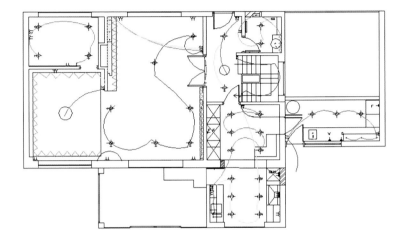

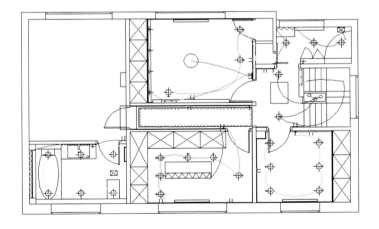

ENSUITE

SHOWER ROOM

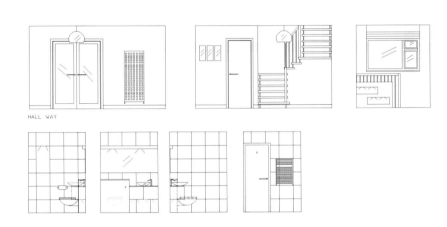

HALL WAY

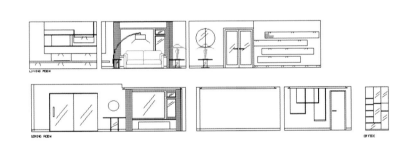

LIVING ROOM

LIVING ROOM

OFFICE

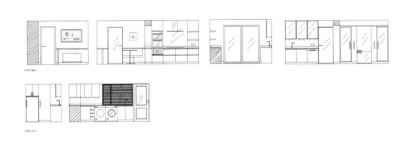

KITCHEN

UTILITY

　　这栋位于英国乡村的大型住宅建于20世纪50年代，房子的主人想对其重新设计，保留原有外观，彻底翻修内部。他们邀请到室内设计师Adrienne Chinn，Adrienne Chinn提出了大规模的翻修计划。拆除原有的旧楼梯，取而代之的是德国进口的玻璃与橡木混搭的开放式楼梯，这样可以使昏暗的走廊变得更加通透。敞开式平面布置的客厅和餐厅是设计的亮点。将原有的隔墙拆除，营造出一个全新而宽敞的空间，Adrienne和她的助手Zhu Cheng（朱橙）在这里用悬臂式的石灰华壁架设计了一个电视主题墙。房子的主人收藏了许多玻璃艺术品，因此，设计师还专门为其打造了一个搁架，用以陈放这些藏品。

　　原来的厨房狭小昏暗，Adrienne建议在朝向花园的一侧安放一道折叠玻璃门，并且专门定做了橡木与玻璃混搭的橱柜。所有的浴室都进行了重新设计，以彩色的杜邦可丽耐人造理石、现代风格的洗面盆、步入式淋浴房以及Jacuzzi水流按摩浴缸为特色。主卧红白相间的设计给人极大的视觉冲击，床头的位置是一面别致的展示墙，上面的格子可以代替床头柜摆放东西。房主夫妇欣然接受了这些设计想法，即刻搬出行李开工翻修。七个月后，大功告成，一个美妙绝伦的设计呈现于眼前！

HOME DÉCOR
Ideas for Interior Space, Function & Color

56th Street Apartment
56号大街公寓

The basic idea for this renovation was to create a straightforward, elegant, and clean design, taking advantage of the luminous and excellent views and creating an atmosphere of comfort, functionality and simplicity.

All spaces were totally redone. Windows that had been blocked over, were reopened, permitting more direct sunlight. Beams and ceilings on the lower floor were leveled to give the area a more spacious look, and also allowing a unified lighting treatment. Hardwood floors were used in the lower floor.

The kitchen was totally redesigned and an opening to the dining room was added, giving this space direct light, ampleness and functionality. In the TV room a closet was transformed into an office space with desk and shelving. In all the apartment the radiators were hidden using in low-built furniture that in some cases were used as shelving. The second floor bathrooms received special attention. Marble and distinctive lighting ensured elegance .The main bath was enlarged to add a shower besides the bath. Wooden planters and deck were added to the terrace, giving this outdoor space an inviting atmosphere.

The overall effect of the apartment is cheerful and sophisticated.

Location: New York, USA
Area: 288m²
Design company: Estudio Ramos
Photographer: Soledad Ramos

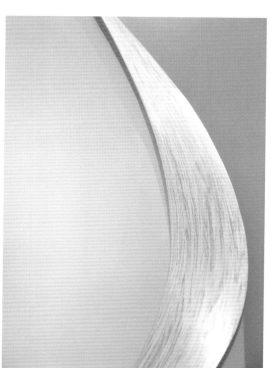

　这个翻修设计的最初构想是以周边绝佳的风景为依托，将其建成一个明快、典雅、洁净的住宅，营造舒适、功能化、简洁的氛围。

　设计对所有空间都进行了改造。一直紧闭的窗户重新打开，使更多的阳光能够照射进来。一楼，水平设计的横梁和天花板给人以更加开阔的视觉体验，同时采用了统一标准的照明系统。地面铺设的是硬木地板。

　厨房被彻底重新设计，添加了通向餐厅的门，使之变得更加明亮宽敞，功能性也更强。客厅的一个小角落被改造成办公区域，配有书桌和搁架。一些低矮的家具遮掩了公寓里所有的散热器，有时，还可以用作置物架。二楼的浴室成为设计的亮点，大理石和独特的灯饰尽显精致优雅。主浴室扩建后，在浴缸旁边增设了一个淋浴区。屋顶平台增加了木质盆栽和露台，让这里的室外空间更具魅力。

　公寓的整体效果可谓品位精致，令人愉悦。

HOME DÉCOR
Ideas for Interior Space, Function & Color

Murray Hill Townhouse
默里山联排别墅

This townhouse is located on the east side of Manhattan in the Murray Hill Neighborhood, New York City, New York, USA. SPG Architects was hired to remedy the spatial fragmentation that had resulted from the numerous renovations. After taking the premises back to base building structure, a new open steel staircase became the chief unifying organizing element for movement and visual focus.

This fully renovated house includes a car garage, living, dining, and kitchen area, a media room, a billiard room, study, sunroom, master bedroom suite and two other bedrooms with baths, plus 3 powder rooms, a penthouse sitting room and two outdoor terraces.

While the house is largely detailed in a minimal and modernist manner, the client requested a house that also references the Moroccan aesthetic that he is especially fond of. SPG Architects designed moments throughout the space that use both traditional and re-interpreted Moroccan motifs, including the custom stone mosaics at the fireplaces, the Spanish Cedar wood doors to the upper level rooms that were handcarved in Morocco, the kitchen cabinetry with wooden screen tracery on the upper cabinets, decorative lighting, and most particularly the screen at the stair, which satisfies railing code requirements while adding a reach pattern that references traditional Moroccan tile mosaics but that are reinterpreted in a laser-cut metal screen with a bronze finish. The details capture the essence of the Moorish influence while still maintaining the modern visual and spatial language inherent in SPG Architect's design philosophy.

Location: New York, USA
Area: 700m²
Designer: Caroline N. Sidnam, Eric A. Gartner
Design company: SPG Architects
Photographer: Daniel Levin

该项目位于美国纽约州纽约市默里山附近的曼哈顿岛东侧。SPG建筑师事务所负责对大量翻新工程所造成的不完整空间进行整合。恢复原有的基本建筑结构之后，一个全新的开放式钢质楼梯成为联合日常活动和视觉效果的主要核心元素。

这栋彻底翻修的别墅包括一个车库、客厅、餐厅、厨房、一个多媒体室、一个台球室、书房、日光浴房、主卧套房，以及两个带有浴室的卧室，另外还有三个盥洗室、一个顶层起居室和两个户外露台。

房屋的设计呈现出一种低调的现代风格，同时，业主还要求设计参照他所钟爱的摩洛哥式的审美情趣。SPG建筑师事务所在整个空间设计中采用了传统的摩洛哥主题图案，并对其进行了重新诠释，包括用于壁炉上的定制的石质马赛克；上层房间的西班牙香柏木门，由摩洛哥工匠亲手雕刻而成；细木板制成的橱柜，上柜的门板为木质格栅，此外，还包括装饰灯具等。尤其是楼梯处的屏风，在满足扶手栏杆规范要求的同时增加出一个延伸模式，参照了传统的摩洛哥马赛克瓷砖，但同时通过激光切割技术对青铜色的金属滤网屏风进行了重新诠释。设计的细节之处吸取了摩尔文化风格的精华，同时保持了SPG建筑师事务所固有的现代视觉和空间语言的设计理念。

HOME DÉCOR
Ideas for Interior Space, Function & Color

Setia Eco Villa
塞提亚生态花园别墅

Derive from the client's brief, the designers decide to play an additive form generator to create a massing and, at the same time, functional space as well. The indoor space was analogized in pure geometric box, which composed, juxtaposed and inter connected one and the other with a "gap" in between, which interpreted to a circulation or buffer zone, to infiltrate natural air and daylight, as much as possible.

In vertical way, the massing were rotated and juxtaposed perpendicular one and the other to create the in between space into roof garden. With this roof garden, the designers try to create a rain water catchments surfaces to replace the ground which covered by the building, also giving an added value to the room at upper level, as a direct outdoor orientation from within. This rotation of the massing also gives a shadow effect to the room below.

The usage of few kind of texture and color to this building, also refer to the richness and diversity of Asian architecture in general.

The usage of wide canopy in the top part of the building also plays a double role, as a place for solar panel, beside as a canopy. This amount of solar panel was calculated to reduce 30 % of energy consumed within this villa.

Location: Kuala Lumpur, Malaysia
Area: 400m²
Designer: Ir. Tonny Wirawan Suriadjaja
Design company: TWS & Partners

根据委托人的想法，设计师决定打造一个集设计概念与功能于一体的生态别墅。室内空间类似于纯粹的几何体并列组成，中间的一个"缺口"将两端相互连接，这里可以被视为一个循环或缓冲区域，能够尽可能地提高采光和通风效果。

以垂直的方式旋转整个设计，竖直地把两端并列起来，将中间的空间打造成屋顶花园。用这个屋顶花园，设计师试图建造一个雨水汇集区域，用以取代建筑物覆盖的地面，同时，作为一个由室内直接通向外部的区域，也增加了上层空间的附加值。这种旋转式设计还为下层的房间提供了阴凉效果。

这栋别墅的设计较少运用花纹与色彩，是因为亚洲建筑设计风格本身的华美与变换。

建筑顶部使用了大面积的遮盖棚，既可以遮阳，同时还能放置太阳能电池板，起到双重功效。这些太阳能电池板的应用可以为整座别墅节约30%的耗能。

HOME DÉCOR
Ideas for Interior Space, Function & Color

Tite Street
泰特街

The house in Tite Street, Chelsea, was created as a personal home for a property developer and his family. Designers from Blacksheep were commissioned to create an entire and integrated interior design scheme for the 4-storey house, after the developer had purchased and then gutted the existing property – building a new front facade in the process-leaving only a bare shell inside.

This blank canvas gave the Blacksheep team the opportunity to create a scheme with a real sense of unity, choosing a colour palette and feel that would unify all the spaces. The scope of the design covered all floor and wall surfaces, interior design, lighting design and interior decoration / styling.

The developer asked for a modern and contemporary interior with clean lines, but once which was easy to live in, with an accent on comfort. Blacksheep linked the spaces by using oak flooring throughout and then created an interior scheme in tones of chocolate, taupe and mushroom to compliment the flooring, using textured materials to create additional interest.

The house is comprised of a gym area on the lower-ground floor, a series of living, dining and reception spaces and five bedrooms, of which three are ensuite. The master bedroom has a separate dressing area lined in oak, accessed through a narrow corridor with padded fabric walls.

Location: London, UK
Designer: Jo Sampson, Tim Mutton
Design company: Blacksheep
Photographer: Gareth Gardner

　该项目位于切尔西的泰特大街，是为一位房地产开发商和他的家人打造的私人住宅。房主买下这栋住宅后，将其已有的设施全部拆除，只留下一个空壳子，试图呈现一个全新的建筑外观。他委托Blacksheep的设计师为这栋四层建筑规划一份整体的室内设计方案。

　这张空白的画布给了Blacksheep团队一个以整合理念为主题的创作机会，选取一种色彩基调，使整个空间和谐统一。整个设计范围包括所有的地面、墙壁、内部设计、灯光设计以及室内装饰的风格和款式。

　房主要求用简洁的线条呈现代风格，并且要舒适宜居。设计师通过全部铺设橡木地板的方式将整个空间串联起来。在内部设计上，以巧克力色、灰褐色及蘑菇色为基调，用以衬托橡木地板，使之和谐完美。同时，也运用了一些带有纹理的材料来增添情趣。

　房子由底层的健身房、一系列的起居室、餐厅、会客室和五个卧室（其中三个配有家具）组成。主卧里，橡木隔断分离出一个梳妆区，可以通过一条墙上贴有织物的狭窄过道进入。

HOME DÉCOR
Ideas for Interior Space, Function & Color

Ojeni Flats
欧杰尼公寓

Lying between one of the chiquest streets in Beyoğlu and the lively heart of Tünel, the Ojeni flats boast stunning views over the whole of historic Istanbul and the Bosphorus. Completely restored and renovated, the flats are designed to suit the busy lifestyle of its residents, with compact features such as kitchens which fold away to present a more clean-cut living space. Oak flooring gives a feeling of warmth, whilst the black-painted bathrooms bring across a sleeker, more modern look. The flats are also furnished with some of Autoban's newest designs including a custom-made version of the luxurious Deco Sofa, the perfect lounging point before hitting the town...

Location: Beyoğlu, Istanbul
Designer: Seyhan Özdemir, Sefer Çağlar
Design company: Autoban
Photographer: Engin Aydeniz

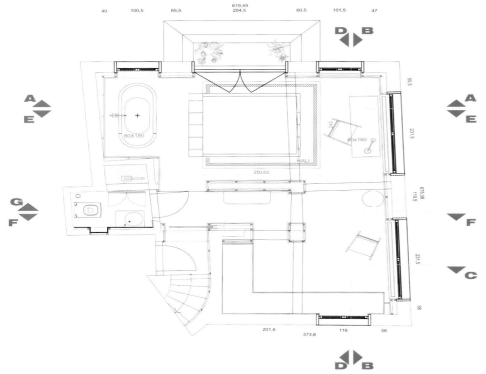

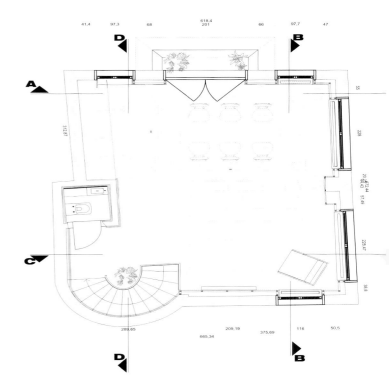

欧杰尼公寓位于Beyoğlu最热闹的街区与繁忙的Tünel地铁中心之间，住在这里可以尽览古老的伊斯坦布尔和Bosphorus的全貌，这令居住者引以为傲。经过彻底的修复和翻新，欧杰尼公寓所呈现出的紧凑空间设计符合了现代繁忙生活方式的需求，例如，将厨房的操作空间隐藏于折叠拉门背后，这样既节省空间，又干净利落。橡木地板给人以温暖舒适的感觉，而以黑色亮色为基调的浴室则更显气派与时尚。此外，还选用了一些Autoban最新设计的家具，包括一款定做的Deco豪华沙发，可以尽享精致的休闲时光……

HOME DÉCOR
Ideas for Interior Space, Function & Color

Buama
布玛住宅

Buama is an interior renovation project for a young couple living in Beykoz, a suburb of Istanbul. As a concept, a new structure is inserted in the existing structure. The new structure maintains the existing heights, yet proposes a more organic solution. The smooth and endless lines reference infinite space, as there is no clear distinction between the floors & walls, or between the walls & ceiling. Light scatters on the organic surface in a way that seems almost limitlessness.

The furniture is selectively chosen to unite many different styles. The young couple has a collection of beautiful antique pieces, and desired to combine the new furniture with this collection. These pieces attract attention in contrast to the smooth surface of the white walls and ceilings.

The proposal for the garden's design also follows the same principles of infinity. Here, the lines seem endless. A place for resting is designed in the space left among the existing trees. This soothing, quiet and relaxing garden space will be where the couple will spend most of their free time.

Location: Turkey, Istanbul
Area: 150m²
Designer: Gokhan Avcioğlu
Design company: GAD Architecture
Photographer: Ozlem Avcioglu

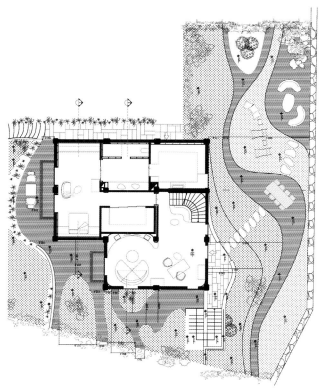

这是一个室内翻修项目，位于伊斯坦布尔市郊的Beykoz，房主是一对年轻夫妇。作为概念，一个全新的结构被引入到原有建筑中。尽管这个结构保留了原有的高度，但却勾画出一个更加系统有机的设计方案。由于地面与墙壁、墙壁与天花板之间没有明显的区分界线，因此，平滑且无限延伸的线条构筑出无限的空间。灯光散落在有机平面上，使其显得无限延伸。

精心选取的家具呈现出不同风格的混搭。这对年轻夫妇收藏了许多精美的古玩，希望能够与新家具完美结合。在白色墙壁和天花板光滑表面的映衬下，这些藏品更显夺目。

花园的设计遵循了同样的无限空间法则。这里，绵延的线条似乎没有边际。现有树木的左侧空间被设计成休息区。主人可以在这个安静轻松的区域充分享受休闲时光。

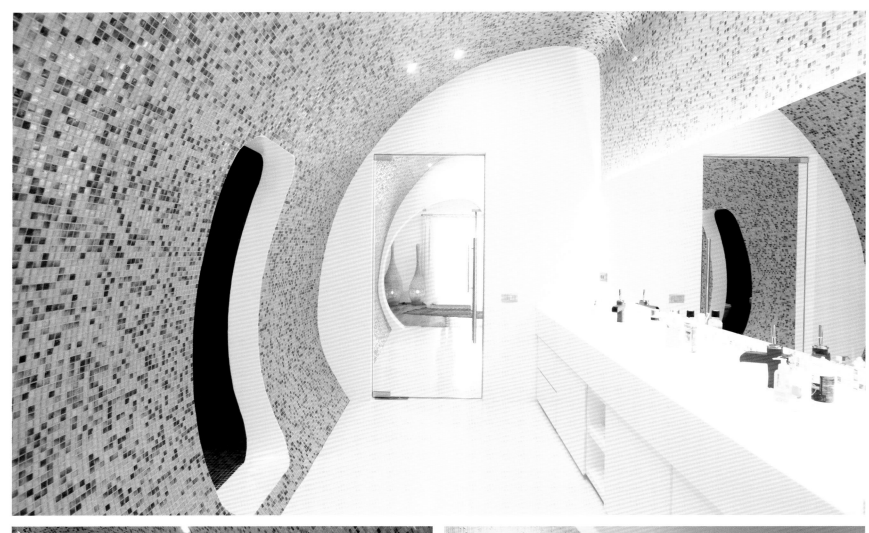

HOME DÉCOR
Ideas for Interior Space, Function & Color

HOME 07
07住宅

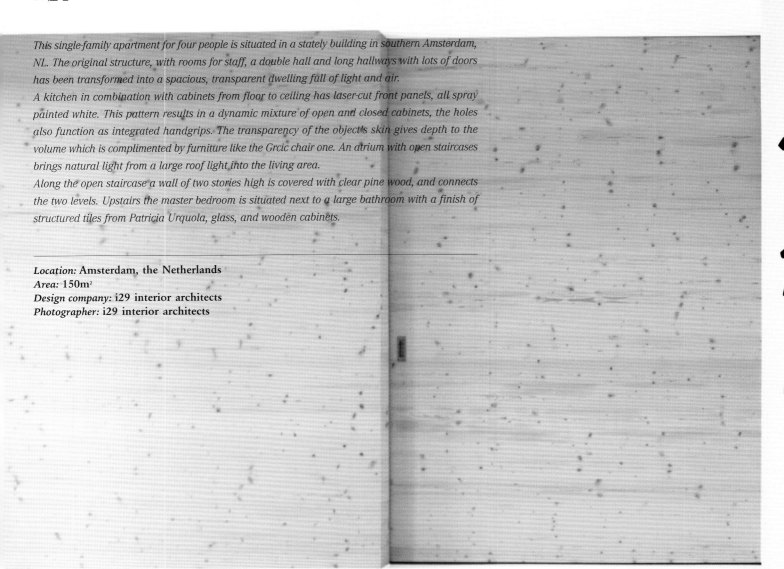

This single-family apartment for four people is situated in a stately building in southern Amsterdam, NL. The original structure, with rooms for staff, a double hall and long hallways with lots of doors has been transformed into a spacious, transparent dwelling full of light and air.
A kitchen in combination with cabinets from floor to ceiling has laser-cut front panels, all spray painted white. This pattern results in a dynamic mixture of open and closed cabinets, the holes also function as integrated handgrips. The transparency of the object's skin gives depth to the volume which is complimented by furniture like the Grcic chair one. An atrium with open staircases brings natural light from a large roof light into the living area.
Along the open staircase a wall of two stories high is covered with clear pine wood, and connects the two levels. Upstairs the master bedroom is situated next to a large bathroom with a finish of structured tiles from Patricia Urquola, glass, and wooden cabinets.

Location: Amsterdam, the Netherlands
Area: 150m²
Design company: i29 interior architects
Photographer: i29 interior architects

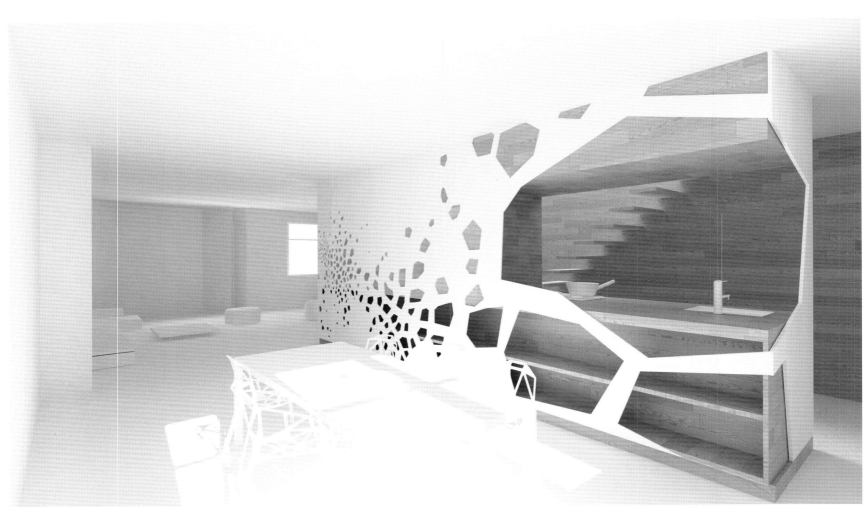
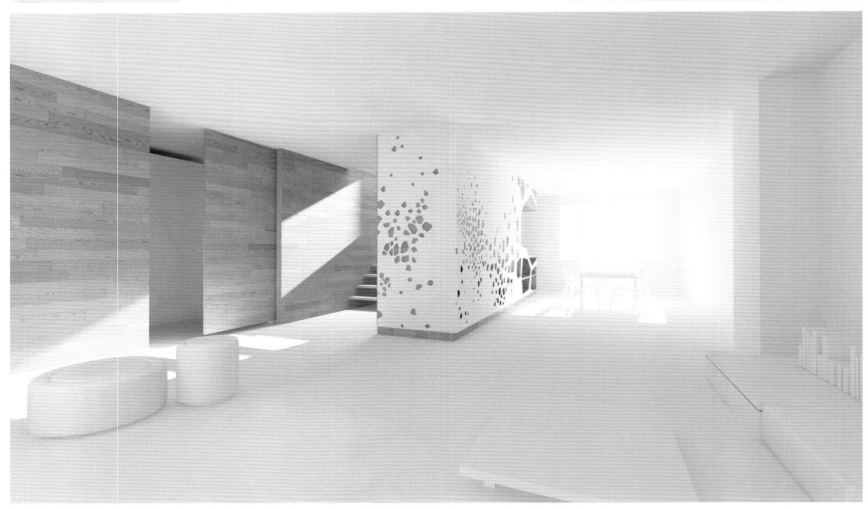

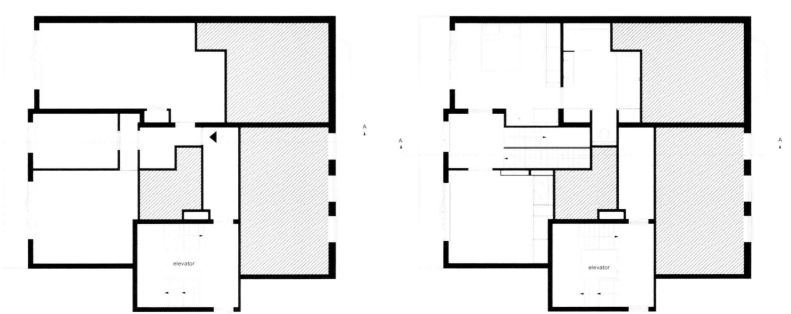

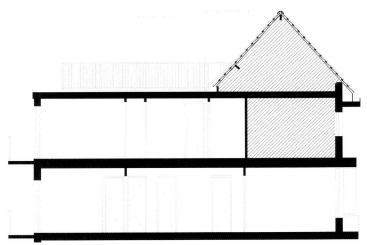
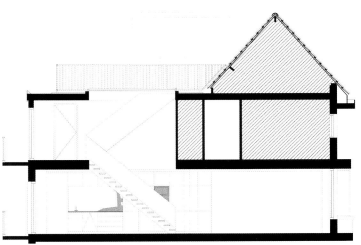

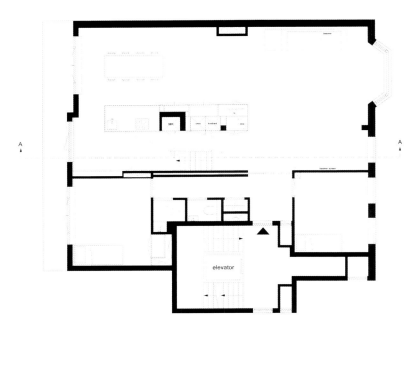

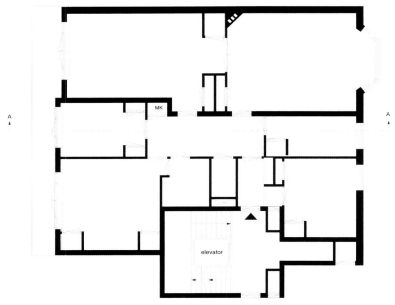

　　这套可供四人居住的单一家庭公寓位于荷兰首都阿姆斯特丹南部的一幢豪华建筑中。所有的房间、双层空间的大厅，有很多扇门的条形走廊……，设计师对这些原有结构进行了改造，针对自然透光与空气流通进行设计，使其更加宽敞、通透。

　　自上而下设计的厨房以一组激光切割板材制作的简约型橱柜为主，后喷以白漆，这样的设计既富有活力，又不乏功能性。橱柜上的空洞起着开阖把手的作用。镂空通透的表面设计，搭配Grcic椅子等家具，加强了整体造型的进深度。中庭的开放式楼梯让自然光透过宽大的天窗直射客厅。

　　沿着开放式楼梯间，覆盖着本色松木的双层高墙壁将上下两层自然衔接。楼上是主卧，配有木质的储物柜，旁边是一间宽敞的浴室，铺设了Patricia Urquola瓷砖。

HOME DÉCOR
Ideas for Interior Space, Function & Color

The Woven Nest
编织的巢

The house assembles around the central open stair, its timber strands growing upwards towards the light and unleashing delicate tendrils to frame each step, a single thin metallic line dancing across their lines to offer the lightest of additional support to the hands that seek it.

To the right, spaces sneak into the stair - as bathroom storage below or the underside of the desk above - while to the right the open treads fan and splay into a generous array of surfaces for the living room. Their lower steps support a seat and soft-spot, while their upper elements flow around the sitter with a sea of books and shelves.

Upstairs, the stair-tree verticals curl into architraves and continue into rooms either side of the eyelid to the sky above. Their lines flow to form a desk and shelving unit in the study, wrapping around to welcome the unfolding sheaves of floor plank that conceal a bed within the floor-depth. The low table/cupboard nestled at the window flows out to form a long courtyard storage bench, which slips back inside as a bathroom counter, carved with a sunken bath. This same surface plunges through the bather's view-slot into the bedroom, a faceted plane (the laundry-lid) folding up to form the final blackout for this bedroom/bathroom opening. It continues as storage into the plinth of the welcoming bed beyond, and onwards as bedside counter before folding back into the wall and the rhythms of the stair beyond.

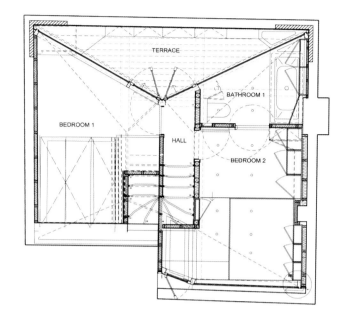

Location: London, UK
Area: 70m²
Design company: Atmos studio
Photographer: Atmos studio, Christoph Bolten

　　房子围绕着中间的开放式楼梯展开，楼梯踏步的木质边缘向上延伸，直至棚顶，自由柔美的曲线框出每一级台阶的造型。作为楼梯扶手，一根纤细的金属杆宛如舞动的线条贯穿其中。

　　右侧，浴室位于楼梯下方的隐蔽空间。开放式的楼梯踏板如同百叶扇一样，朝向客厅一面展开，形成一组开放式的搁架，较高的几层可以当书架摆放书籍，而人们则可以坐在较低的位置尽情阅读。

　　往楼上看，楼梯宛如一棵大树蜿蜒向上伸展，直至顶梁，接着延伸至两边卧室的尽头。这些线条蜿蜒流动，构成书房的书桌和书架，延伸的楼梯台阶环绕出一层层格子架，每一层都能容下一个人躺在里面。窗边，低矮的桌子和橱柜的线条延伸至屋外，形成院子里的一条长凳，另一端通向浴室，作为嵌入式浴盆的柜体。从浴室的角度可以看到这些木质线条继续延伸，形成卧室区域，线条的立面围合出卧室的床。窗边与墙面的凹陷部分可作为床头的收纳区使用。整个楼梯的设计宛若一篇灵动的乐章。

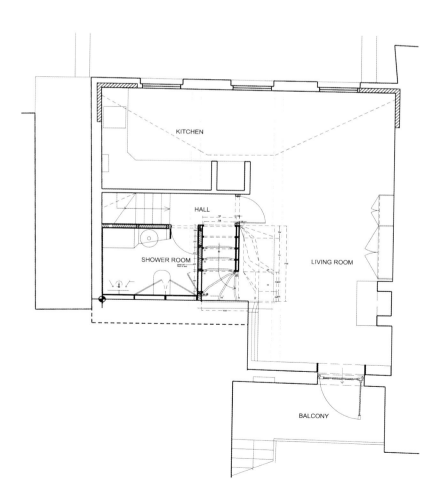

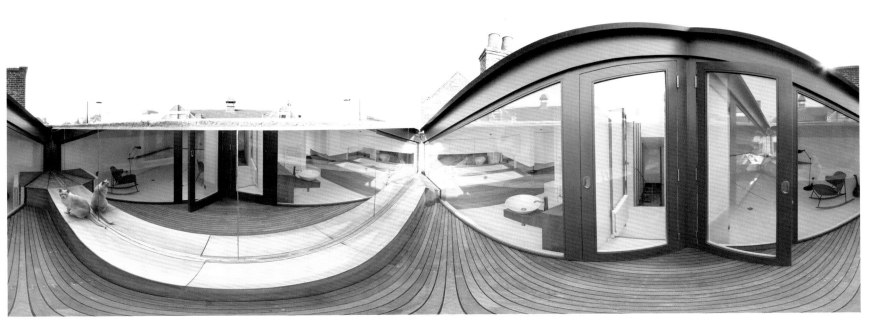
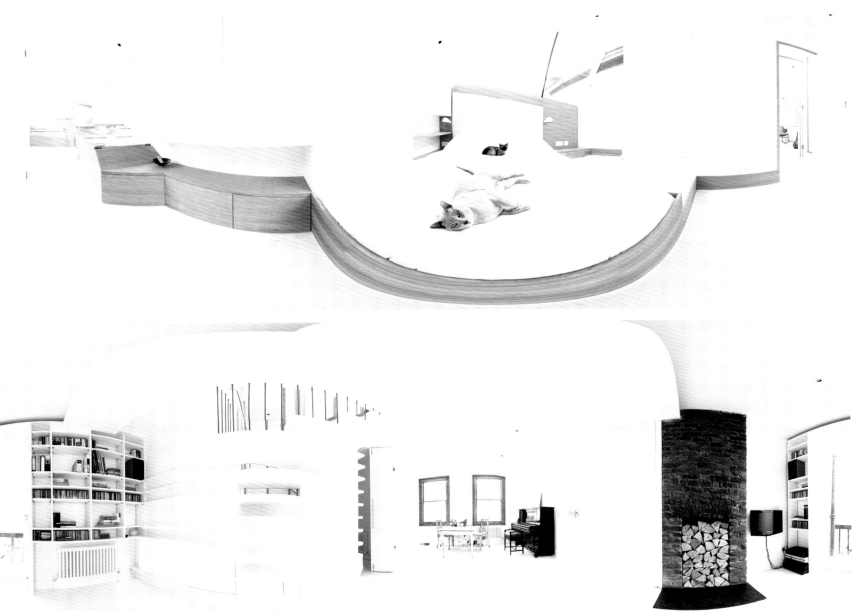

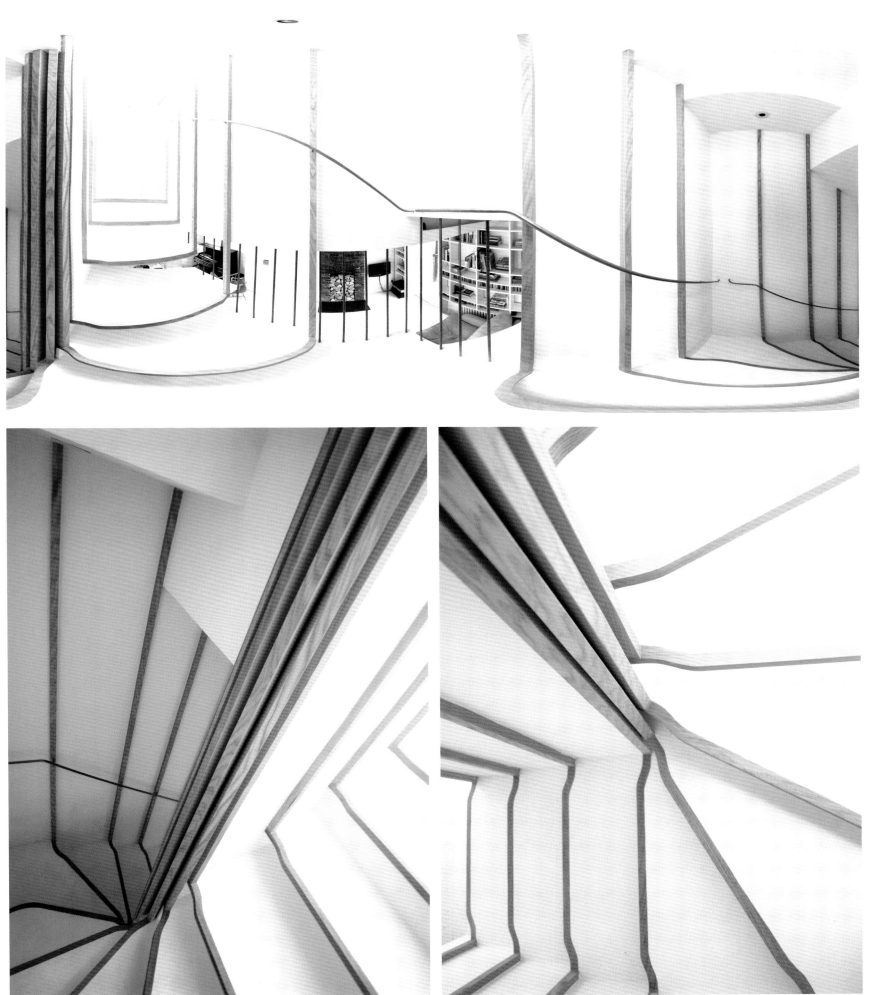

HOME DÉCOR
Ideas for Interior Space, Function & Color

Greenwich Village Penthouse
格林威治村顶层公寓

This renovation transforms what was a beautiful, but austerely minimalist one-bedroom penthouse into a more livable apartment without losing its appeal as a glowing modern space high above Greenwich Village.

Each floor incorporates a continuous visual plane running north-south, which separates the core private functions from the more public living spaces. Parallel translucent panels can be manipulated to create levels of separation at the openings in these walls. Silvery wood paneled boxes contain the kitchen upstairs and a second bedroom with a thickened wall containing various media and storage requirements downstairs.

The warm natural tones of the furnishings contrast nicely with the abstract neutral architectural palette. The palette of this SPG-designed apartment consists of white, off-white, and soft grey architectural surfaces that are in turn complemented by warm golden and natural tones of the furniture and furnishings. The carefully determined off-white of the walls glows and changes over the course of the day and evening depending on the quality of light.

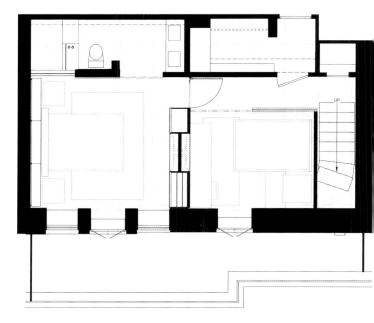

Location: New York, USA
Area: 111m²
Designer: Caroline N. Sidnam, Eric A. Gartner
Design company: SPG Architects
Photographer: Daniel Levin

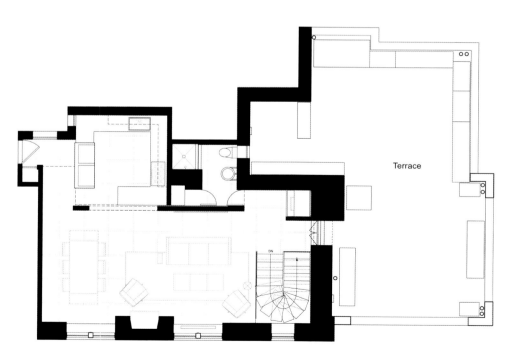

这是一个建筑改造项目,将一个原本漂亮、但却极为简朴的一居室顶层公寓改造成一个不失魅力、更加宜居、且更具现代特色的居住空间。

每一层都包含一个连续的南北走向的视觉平面,将主要的私密空间与公共的居住空间进行分隔。可平行移动的半透明嵌入门板将空间隔离区分。银色的木质墙板将楼上的厨房、次卧以及楼下的储藏室包裹起来。次卧还采用了能够容纳多媒体的加厚墙壁。

室内陈设的格调温馨自然,与抽象的中性建筑色彩恰好形成对比。SPG建筑师事务所设计的公寓建筑表面通常为白色、灰白色,或是浅灰色,而家具及室内陈设则以温暖的金色和天然本色作为补充。白天和晚上的不同光线下,精心打造的灰白色的墙壁也随之富于变化。

HOME DÉCOR
Ideas for Interior Space, Function & Color

Shelter Island House
牛尾洲住宅

The client likes to entertain, have guests out all year round, and I love to cook. So, they wanted to have a place they could enjoy with family and friends. The house and grounds were neglected for many years and had no connection to the outside for use of the yard, etc. Actually, the yard itself was unusable. There was only one bathroom upstairs and two bedrooms, but the client did not want to change the appearance and did a large add on, or alter the shape. The interior had many misdirected renovations over the years, doors built into the floor that could not open, tiled ceilings, and the house was sinking on rotting wood posts. Connections were made to the exterior with new French doors off the dining room, a freestanding pergola added, a 17'-0" freestanding reclaimed brick fireplace constructed at the end of the pergola. This created an expanded living room/dining room area at the exterior, but also helped to balance the house on the site. A custom pool with long cocktail steps was added with an open area for laps. The arage/tool shed was converted into a pool house which added an additional 28 sq.m. The entire site was rethought, so that the entire property could be used in different ways and in different seasons. The designer even makes fires at the exterior fireplace in the winter, since the client can see it from the dining area and if not too cold to sit by the fire after dinner. The first floor of the house was gutted, new footings to stop the house from sinking, new custom designed cabinetry, new wide plank floors, and a new bathroom added. The new 1st floor bathroom has a door to the exterior for easy access from the pool. The designer lightened up the house with whites, off whites and sand colors. Then used black and nature wood as contrast. There are modern pieces, antiques, and found/altered objects. The new bathroom has a steel framed glass partition.

Location: New York, USA
Area: 130m²
Design company: Schappacher White Ltd
Photographer: Laura Moss

该项目的委托人十分好客，常年有宾客拜访，并喜欢亲自下厨。因此，他们希望有足够的空间来款待亲朋好友。这栋房子和庭院已经多年没人照看，与户外相连的院子也没有充分利用。事实上，这个院子本身毫无用处。楼上仅有一间浴室和两间卧室，但是客户并不希望更改外观并做一些搭建，或是改变建筑的外形。多年来，都只是对其内部进行一些不适当的翻新改造，嵌入地板的门无法打开，屋顶天花板使用瓦管排水，房屋的木桩基础已经腐化，开始下沉。新设计的餐厅采用全新的法式落地门与户外相连，添设了独立花藤，花藤末端用可再生砖搭起一个17'—0"的独立壁炉。这样在户外打造了一个延伸的客厅或是餐厅区域，并且有利于保持房屋的平衡。露天区域添加了一个定制设计的泳池，与狭长的混合阶梯搭接，同时把车库（工具房）改造为泳池室，格外增加了28平方米的面积。设计师重新考虑整个空间，以便可以通过不同方式，在不同季节充分利用空间。他甚至在冬天用户外壁炉生火，主人坐在餐厅便可观赏到这番景致。如果天气不太冷，还可以在餐后围坐炉边烤火。设计彻底推翻了原有建筑一楼的布局，新的地基可以防止房屋沉降。此外，还添加了全新定制设计的细工家具、宽条的厚木地板和一间崭新的浴室。浴室的门朝向户外，方便进出泳池。设计师使用白色、米色和沙色来提亮房间，同时运用黑色和原木色作衬托。室内摆设了许多现代家居摆件、古董、自然艺术品或是改造的旧物。新浴室还采用了钢架玻璃隔板。

HOME DÉCOR
Ideas for Interior Space, Function & Color

Attico
阁楼

The idea of the project was to expand the space of an attic which, despite being quite large with its 240 square meters, appeared cramped by a pitched roof that was compressing its volume. The final result of the conversation is a bright, spacious and dynamic apartment that perfectly reflects the owner's needs.

Converting this kind of apartments can be quite difficult because the attics have very little room at the roof's ridge and require customized spaces. The practise Damilano Studio took the challenge, managing to create large rooms with few dividing walls, to give more breath to the apartment.

By knocking down the portion of the wall near the main door, a window on the cut-away roof terrace could be opened on the ceiling. This skylight floods with light, and a grand piano immediately catches the eye of those who enter the apartment. The sleeping area is designed as an independent area, an alcove where to retire at the end of the day. The bed and the bathtub, separated by a glass, create two spaces for the wellbeing in communication with each other and the soft light dampens the rigid geometry of the room. The bathroom is the only closed room, where the shower is obtained at the cockloft level to exploit its height.

The chosen colours and materials help giving breath to the spaces and the white reflective floors create homogeneous surfaces next to the parts made of chestnut wood and facing in Indian stone.

Location: Cuneo, Piemonte, Italy
Area: 240m²
Designer: Duilio Damilano, Claudia Allinio
Design company: Damilano Studio Architects
Construction: Arch. Massimo Aimar
Photographer: Andrea Martiradonna

该设计的意图是要扩大阁楼空间。尽管整套住宅的面积为240平方米，但阁楼斜屋顶的压迫感使得空间颇为局促。设计最终向我们展现了一个明亮宽敞、充满活力的房间，满足了主人的需求。

实现这一转变很有难度，因为阁楼屋脊两侧的空间十分狭小，得按需定制空间设计。Damilano工作室接受了挑战，采用无隔墙设计，尽量加大空间感，让房间更加通透。

将正门附近的部分墙体拆除，在屋顶的斜坡处开设天窗。阳光透过天窗洒向屋内，使得一架大三角钢琴立刻成为最抢眼的亮点。

睡眠区域被设计为一个独立的凹进空间，供主人休息。床和浴缸之间使用了玻璃隔断，营造出两个相得益彰的空间。柔和的光线削弱了房间死板的几何结构。浴室是这里唯一的封闭空间，淋浴器充分利用了这里的空间高度。

设计选用的颜色和材料也让空间变得明亮通透，白色的光亮地面与采用印度石贴面加以修饰的栗色木地板和谐呼应。

HOME DÉCOR
Ideas for Interior Space, Function & Color

Penthouse Serrano
塞拉诺顶层公寓

In this project the designer solved the usual separations of spaces inside apartment (flat and orthogonal walls) introduction troncocónicos elements that go from the floor to ceiling. These elements try to impose their strong volumetric, acquiring the range of massive objects, individualized, therefore with their own personality. By perceiving these objects as interposed pieces we obtain a sensation of continuity, as the visitor moves through the house and never sees the end of each volume and really never preserving the totality of the house. This gives him the sensation that the house is much bigger than it really is. These curved volumes have the ability to lead the light through all the spaces of the house; by provoking rebounds like a fluid that goes throw the house, they really flood the house with light, illuminating each interior space. These forms do not try to shelter or contain any space; they have not been designed to shelter places of permanency or static rooms. They are surfaces that reinforce the internal movements of the house, not only of the visitor but of the light, walking through the space we can appreciate a continuous change of form and light, with the contrast of a soft perspective which is at the same time powerful. This contrast is the one that can offer a continuous life inside a space. This geometric freedom, offered by the curve, allows taking the maximum advantage of the m² of the floor plant and turns this house into a place without any dead spaces or dead end.

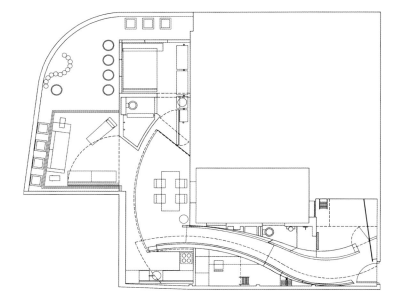

Location: **Madrid, Spain**
Area: **200m²**
Designer: **Hector Ruiz-Velazquez**
Design company: **Garcia Ruiz architects**

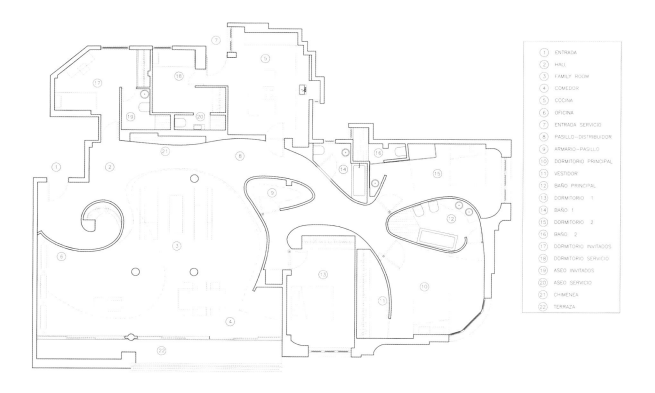

在该项目中，设计师拆除了公寓内部空间的常用隔断（即单元之间的正交墙），在地面和天花板之间采用troncocónicos设计元素。这些元素试图利用其庞大的体积，形成不断延伸的大块柱体，以此来彰显个性。通过这些嵌入的柱体，人们可以获得感官上的连续性。客人在房间里走动时，根本无法看到每个空间的尽头，也觉察不到房间的整体，这样就使得房子看起来比它的实际面积要大得多。这些弯曲的柱体结构可以引导光线照亮屋内的所有空间，灯光犹如流淌的液体，通过反射射入房间，照亮每个角落。这种设计形态没有试图遮挡或是隐藏任何空间，也没有被用于遮挡那些永久性或是固定不变的空间，而是用于加强内部空间的动感，无论是人还是灯光，游走于这一空间，可以欣赏到形态和灯光的连续变化，既柔美，又富有强烈的视觉冲击。而这种对比也可以赋予空间内部连续的生命力。这种通过弯曲构造形成的自由几何形态使房屋的面积最大化，并将这里变成一个没有任何死角和尽头的空间。

HOME DÉCOR
Ideas for Interior Space, Function & Color

Twin Loft
双子座阁楼

A single-cell organism divided into 2 volumes structurally identical and internally "mirrored". This is the project for the restructuring and Interiors of twin lofts designed by Federico Delrosso in the post-industrial Milan: the work turned on a former factory, once an industrial suburb, today a residential district.
"They have the same design as the internal spaces, because they were created by dividing a single sleeve-shaped space lengthwise. But starting from the personal data of the residents (one is Delrosso's own studio-home, the other is the private home of Alessandro Sartori, creative director of Zegna's Z line) I devised a project which developed the architectural details, the furnishings... the experience of the home in an individual way, in short by using materials and colours like the letters of an alphabet: to write similar or opposed phrases, but always with a coherent grammar."

Location: Milano, Italy
Design company: Federico Delrosso Architects
Photographer: Matteo Piazza

　　将一个空间一分为二，隔成两个内部结构相同的"镜像"——双子座阁楼。该项目是对后工业时代米兰设计师费德里克·德尔罗梭设计的"双子座阁楼"的重新布局与室内装饰：将一个位于工业郊区的旧厂房变成今天的这个住宅小区。

　　"它们具有相同的内部空间设计，因为它们是由一个独立的筒形空间纵向隔开而形成的。然而，根据住户的实际情况（其中一个是Delrosso自己的工作室兼住宅，另一个是杰尼亚公司Z系列的创意总监亚历山德罗·萨托利的私人住宅），我设计了这个项目，发掘建筑的细节、配套的家具……基于个人对'家'的独特体验，简而言之，利用类似于字母表中字母的材料和颜色；写上意思类似的或相反的短语，但这些短语的语法却是连贯一致的。"

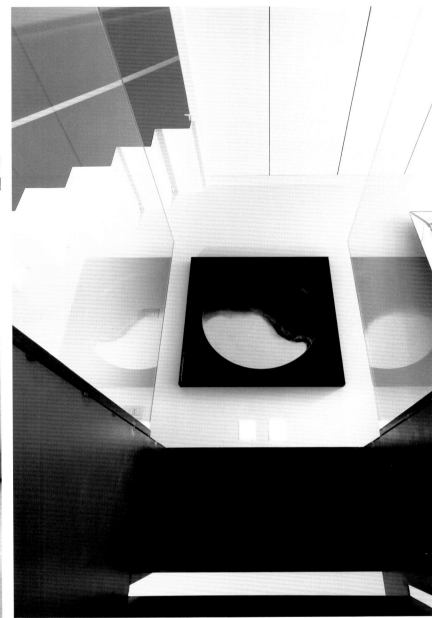

TWINS LOFT - MILANO

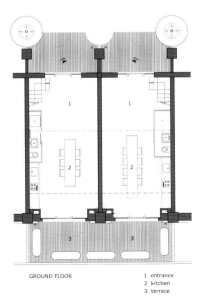
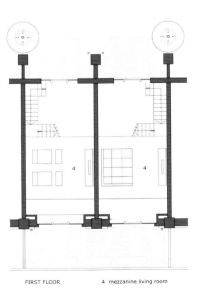
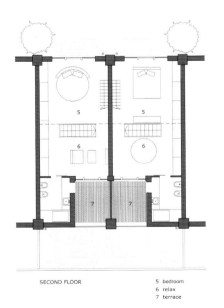
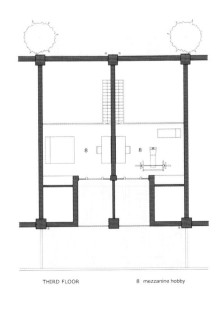

GROUND FLOOR
1 entrance
2 kitchen
3 terrace

FIRST FLOOR
4 mezzanine living room

SECOND FLOOR
5 bedroom
6 relax
7 terrace

THIRD FLOOR
8 mezzanine hobby

HOME DÉCOR
Ideas for Interior Space, Function & Color

Istanbul Suites
伊斯坦布尔套房

Istanbul Suites is located in the Cihangir neighbourhood which is very close to the historical Istanbul texture. Through the windows, you can view all of the historic quarter, including Topkapi Palace, Galata Tower and the big mosques in Sultanahmet. Keeping a modernist approach, the sandstone facade of the building was inspired by the strong appearance of the Tophane-i Amire Building at the end of the street. The long windows and French balconies of Istanbul Suites reflect the architectural style of the Italian Hospital just opposite the hotel.

Inside the lobby and rooms, visitors may join together, allowing the local Cihangir daily life to carry on. Each room was designed as one space where you can eat, rest and sleep. Patterns made from laser cut wood or iron create layers, and combine with existing Autoban furniture. Some of the furniture has been adapted to the hotel concept. The bathroom walls are made of traditional Marmara marble which is the main material in the Ottoman architectural style. The whole ambiance provides a unique experience where the visitors can broaden their horizons and breathe Istanbul.

Location: Cihangir, Istanbul
Designer: Seyhan Özdemir, Sefer Çağlar
Design company: Autoban
Photographer: Ali Bekman

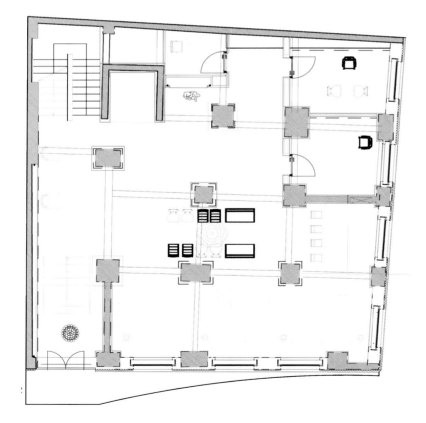

该项目位于有着悠久伊斯坦布尔历史特色的Cihangir街区。透过酒店的窗户，你可以欣赏到每一处有着悠久历史的景观，包括Topkapi Palace宫殿、Galata Tower塔以及Sultanahmet的大清真寺。酒店的设计保留了现代主义手法，建筑物的正面采用砂岩为材料，其灵感得益于位于街道尽头、有着坚固外观的Tophane-i Amire文化中心。长形的窗户和法式阳台则体现了与酒店风格相反的意大利医院的建筑风格。

在酒店的大厅和客房，来宾可以聚在一起，体验Cihangir当地人的日常生活。每个房间都可以就餐、休息、睡觉。室内装饰图案的材料是由激光削切的木头或铁皮制成，家具由Autoban工作室设计。其中的一些家具已经被改造，以便适应酒店的使用。浴室的墙壁由传统的Marmara理石砌成，它是Ottoman建筑风格的主要建筑材料。整个设计格调独一无二，来宾可以在此开阔眼界，品味伊斯坦布尔的气息。

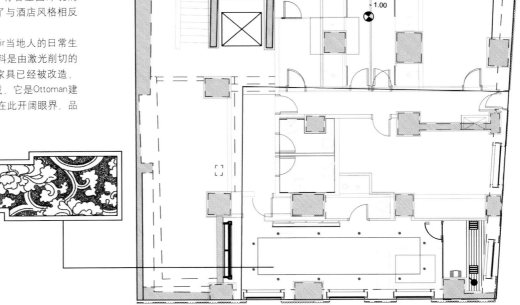

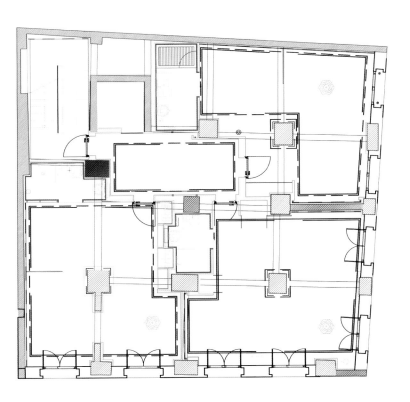
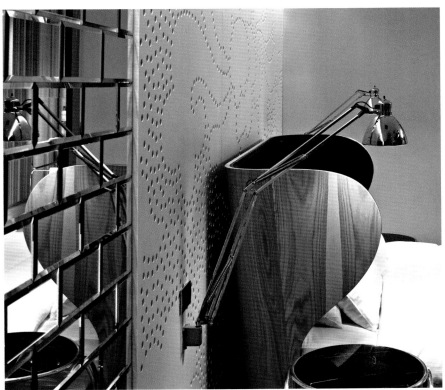

HOME DÉCOR
Ideas for Interior Space, Function & Color

35 White Street
怀特街35号

This loft in the heart of Tribeca was the painting studio of abstract expressionist Barnett Newman. Daylight from thirty meters feet of windows fills the space. Our intention to preserve and enhance the essential character of this historic loft - while crafting within it a domestic space for a family with a young child - relied on a sense of openness and light. Operable panels of prismatic glass configure to create either dramatic openness or total privacy. Custom mahogany and bronze panels enhance the natural breadth and depth of the loft. A state-of-the-art cook's kitchen and French limestone bathrooms make this space equally suited for dramatic entertaining and intimate daily living.

这个阁楼项目位于特里贝克地区的中心，是抽象表现派画家巴奈特·纽曼的工作室。白天，阳光从30平方米高的窗户洒满整个空间。它是一座具有历史意义的建筑，设计初衷是在基于开放性和通透性的基础上，对原有建筑的基本特征予以保留和改进，将其打造成可供一个有小孩的家庭使用的居住空间。棱镜玻璃操作台的安放将引人注目的开放空间与整个私密空间分隔开。定做的红木青铜色面板更增加了建筑的空间感。最先进的厨房设施以及法式石灰岩浴室使得这里同样适合日常生活与休闲娱乐。

Location: New York, USA
Area: 177m²
Designer: Andre Kikoski Architect
Design company: Andre Kikoski Architect
Photographer: Peter Aaron at ESTO

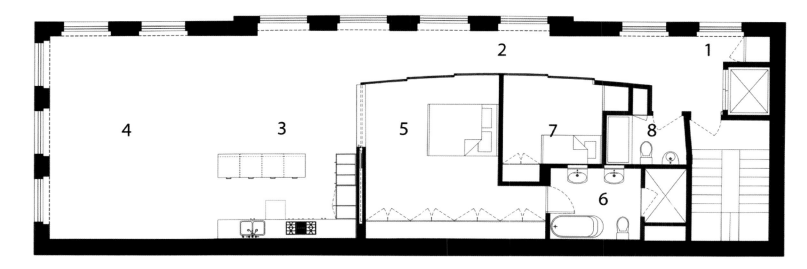

1 - FOYER
2 - GALLERY
3 - DINING ROOM
4 - LIVING ROOM
5 - KITCHEN
6 - MASTER BEDROOM
7 - MASTER BATHROOM
8 - BEDROOM 2
9 - BATHROOM 2

Index | 索 引

A-cero Joaquin Torres architects

A-cero Joaquin Torres architects, architecture and town planning studio, is a company created in 1996 vowed to the integral development of architectural projects. Its growing clients' demand has recently forced it to enlarge its field of action up to work management, carrying out deliveries with vacant possession of works previously designed by it. Throughout its professional evolution, it has maintained different collaborations with building and property development companies, applying for competitions or performing consultancy tasks. Its working method is based on the detailed analysis of the client's needs program. All the dimensions of a problem are posted, all the possible solutions to the problems are studied, and all the material possibilities of building the solved problems are analyzed. The materials and the details are developing and persuading. During this process, architecture appears effortlessly.

Alex Haw

Alex Haw is an architect and artist operating at the intersection of design, research, art and the urban environment. He runs atmos, a collaborative experimental practice which produces a range of architecture and events including private houses and public buildings, videos, installations and larger public commissions. Projects include Sunlands, Hurry Up Please It's TIME, Beammobile, etc. LightHive, a 3d-cctv transmutation of everyone in the Architectural Association into light, was highly commended at both the FX and Lighting Design awards; Work/Space/Ply/Time, a CNC-plywood rapid-assembly pavilion, won the Urbantine competition and is currently touring China with British Council funding. Alex has taught design studios at the Architectural Association, Cambridge University and TU Vienna. He has written for AAFiles, Blueprint, Contemporary, Building Design and the Architect's Journal.

Adrienne Chinn Design Company Limited

Adrienne Chinn Design Company Limited was formed in 1999 with the intention of bringing high quality original interior design to the residential and commercial marketplaces. The success of the company is built on the foundation of Adrienne's Honors Diploma gained in London and the subsequent practical experience she gained working alongside one of Chelsea's most successful domestic interior designers. The company has a large portfolio of completed projects ranging from single room renovations, through whole dwelling refurbishments up to international commercial projects. Adrienne attempts to educate as well as serve. She does this by way of publishing interior design articles on the nitty gritty of interior design, by working closely with her former college to plough back the experience she has gained, by participating in lecture events and above all by maintaining an open policy about her interior design fees and working practices.

Kelly Hoppen

Kelly Hoppen is best known as the interior designer whose calm, elegant aesthetics has permeated our consciousness and achieved an iconic (and much imitated) status.
Having started her business at the age of only 16, Kelly is able to look back at a vast amount of experience and has established herself as one of the best know International Interior Designers. Her celebrated interior design studio, which employs 35 members of staff who manage projects around the world, reaching from 0.5 million to 8 million pounds has completed international schemes for houses, apartments, yachts, ski chalets, a sports centre, hotels and numerous corporate spaces, private jets as well as the interiors for British Airway's first class cabins. Kelly has also successfully implemented her unique approach across a number of business areas, firmly establishing her reputation as designer, retailer, author, educator, innovator and inspiration. Her books have been translated into numerous foreign languages and her work has been on the front covers of magazines worldwide.

Architecture in Formation

Architecture in Formation PC is a young, emerging New York City architecture and design firm founded in 2001 by Matthew Bremer, AIA. The practice is committed to "making" architecture with ever-fresh conceptual rigor and formal dexterity. A deep understanding of the relationship between a project and its larger physical and cultural context informs the firm's methodology: expansive in-depth research at the service of thoughtful and innovative objects, buildings, spaces, and places.
AiF is committed to solving environmental problems using equal parts prudence and audacity: a constantly watchful eye, and inordinate blind faith. They are yesterday's children and tomorrow's mentors. A given project is using appropriate technologies at the service of environmental circumstance.

Andre Kikoski Architect

Andre Kikoski Architect is a Manhattan-based multi-disciplinary design firm that is committed to artistic innovation regardless of budget, genre or client challenge. Our passion for material research, our detail-orientation, and our client-centric approach have won the firm clients in a wide range of categories – from arts and culture to hospitality, from high-end residential to commercial real estate.
Andre Kikoski Architect is the recipient of eleven international design awards and nominations.
Andre Kikoski is a member of the Executive Committee of the Alumni Council of Harvard University's Graduate School of Design; the Harvard Alumni in Real Estate; and the Thomas Moran Trust in East Hampton, New York.

TWS & Partners

TWS & Partners was established in 1998 in Indonesia. It is led by Tonny Wirawan Suriadjaja.
Tonny Wirawan Suriadjaja , born in 30 January 1972, Jakarta, Indonesia , is an architect who always tries to find any innovation in architectural and design interior.
May 1995 - July 1995 work with Gunawan Tjahjono Ph.d to design National Museum in Korea
July 1995 - November 1995 work with Shimizu Lampiri Consultant at Jakarta as an Architect
November 1995 - June 1998 work with Ciputra Development at Jakarta as an Architect
June 1998 - Present establish TWS & Partners Design Team

Damilano Studio

This practise of architecture is named after its founder Duilio Damilano who moved from Polytechnic of Turin to Milan, in order to follow a work-shop by Daniel Libeskind, where he's developed his concept of architectural research directed towards the concreteness of space.

His passion for volumes, instead, comes from a family of sculptors. His father and his brother have, in fact, passed on to him an interest for the plastic and material aspect of every sculpture or architecture. Duilio Damilano says that he's always been attracted by architectures since childhood. His design path begins from the study of the light and how this affects and moulds the shapes.

"It is an instinctive vocation – says Duilio Damilano – a feeling that starts from deep inside and only at a later time is conveyed inside the channels of rationality. Certainly is something that's always been part of my approach to architecture and which goes into my entire project. I mean the kind of lighting that goes beyond the physical aspect of the luminous element to become an integral part of the architectural design. A project can't peter out with the daylight".

studioMDA

Markus founded studioMDA in New York 2002. studioMDA's approach is to creatively engage all parties at the very beginning, forming an inspirational collaboration in defining architecture. Calling a wide range of consultants beyond the architect's typical collaborative, studioMDA has teamed with artists, video artists, and choreographers, among others to define space generated by various artistic endeavors. Within this context, architectural syntax can become layered into micro and macro scales, balancing the smallest detail with the topographical/urban fabric. studioMDA's awards include: In 2006 it was selected as one of 5 finalist designers for MOCA Cleveland's new building. In 2007, it was awarded the First Prize to design and develop the Brooklyn Arts Tower. In 2009, it was awarded the First Prize for the Center For Advanced Mobility in Aachen, Germany as Architect and Project Manager.

Concrete Architectural Associates

Concrete Architectural Associates is founded in 1997. The present director of Concrete Architectural Associates, Rob Wagemans, was born in Eindhoven on 13th of February 1973, he is a Master of Architecture Utrecht.
Erik van Dillen (at this moment he is just creatively involved with Concrete), interior architect, was born in de Bilt on 27th of April 1960, catering industry skills in the kitchen, painting restorer.
Concrete originally was founded by Rob Wagemans, Gilian Schrofer and Erik van Dillen. They met each other by a not realized project, a head office in Amsterdam for Cirque du Soleil. Gilian Schrofer left Concrete in 2004 to start his own company.
Concrete Reinforced is founded in 2006 by Rob Wagemans and Erikjan Vermeulen (present co-director of Concrete Reinforced). Erikjan Vermeulen is a Master of Architecture. He worked for different architects to start his own company in 2003.

PARASITE Studio

PARASITE signifies for us a change of attitude. It means to refer to the project from "above". We are interested in the conflict between old and new and not in the integration of the design through materials and details. We believe that exactly this conflict could generate a different type of spatiality. We are not interested in the forms of "correct architecture" because we want to make a further leap. Architecture has to start to shape space in a different way, details must be seen in a contemporary fashion, and they have to complete space instead of being the generators for architecture and theme. The lack of details of the contemporary buildings must be understood as a change of architectural attitude, architecture has not grown poorer but has become something different.

Sebastian Knorr

Sebastian Knorr (43), together with Heiko Ostmann and Moritz Knorr, is the founding partner of tecDESIGN in the United States and tecARCHITECTURE in Switzerland, and tecHONG KONG. He is the head of design for the combined companies and the CEO of tec in Los Angeles.
In 2000, after winning the International Design Competition for Pier40 in New York, tecARCHITECTURE moved from New York to Los Angeles. Since then tec has built, and is currently working on, multiple national and international projects on various scales, in Europe, the United States and Asia.
Sebastian has won various national and international competitions and awards. He has been the lead designer on tec's various international projects. tec's expertise and experience, having researched and practiced sustainable design since the early 1990's connects the various projects and recently led to "ECO-CITY" comissions in Europe and Asia, with a special emphasis on climate neutrality.

Global Architectural Development

Global Architectural Development is an Istanbul and New York based company, which performs architectural practice, research and concept design since 1994 owned by Gokhan Avcioglu and GAD's global collaborators.
Contemporary and current architecture, urbanism, software, consumer habits and behaviors and approaching to the projects holistically are among its field of interest. GAD understands architecture as a practice that relies on the experiment, and values historical precedents and new ways to combine both in a mutually benefiting fashion.
GAD Architecture is committed to finding innovative approaches to architecture and creating new spatial experiences with projects and ideas.
GAD has won numerous awards including the 1997 Turkish Architecture Prize for the design of a Public Park in Istanbul and the 2001 Cimsa Design Prize for outdoor seating, a bronze medal in Miami Biennale for Borusan Exhibition Center in 2003.

Simone Micheli

Simone Micheli is a university contract professor. He founded the architectural studio with his name in 1990 and the Design Company "Simone Micheli Architectural Hero" in 2003. His works concerning architecture, contract, interior design, exhibit design, design, graphic and communication are strictly linked to the sensorial glorification. He is the curator of some experimental events for some of the most qualified international fairs. He shows his projects in the most important worldwide architecture and design exhibitions. He is the curator of some thematic "contract" exhibitions in the most important international fairs of this field. He represented Italian interior design taking part to the XXX Congreso Colombiano de Arquitectura in Baranquilla and participating at the International Conference of Architecture in Hanover; in 2008 he has signed the event called "La casa italiana" for ICE and Fiera Verona in collaboration with Abitare il Tempo-Acropoli, at the MuBE (Brazilian Museum of Sculpture) in San Paolo.

SPG Architects

SPG Architects is a mid-sized architectural firm based in New York City. Headed by two partners, SPG is a strong design practice that also provides a comprehensive range of architectural and interior services. The firm has served a wide variety of clients with a range of regional, national, and international projects that include numerous co-op and condominium apartments, urban townhouses, second homes, free-standing houses and residential compounds, corporate offices and retail establishments.
SPG Architects' modernist approach to design allows for the various functions of a space to be organized and expressed, while eliminating the cacophony of the untended environment. Architectural ideas are drawn from the project site and the client's needs and desires. These then are expressed through manipulations of form and light. An interest in up-to-date building technologies and construction materials, both natural and man-made, further informs the design.

IN SITU

IN SITU is a Bucharest based practice committed to architectural and interior design. The office was founded in 2006 by architect Ion Popusoi with experience based on a large amount and variety of projects and competitions in Romania and abroad. For each project we are motivated to design memorable, sustainable and efficient solutions that add value to the inhabitants' lives, employing a versatility and passion that transcends scale and budget.

D'Aquino Monaco Inc

Designer Carl D'Aquino and partner architect Francine Monaco along with their firm D'Aquino Monaco approach every project with a unique artistic vision encouraging a strong personal dialogue with their clients. They have the ability to transform and create space architecturally as well as decoratively right through to the details of furniture, fabrics, and lighting to create a unique coordinated vision. Each project is a true representation of the collaboration between client and designer. D'Aquino Monaco lists an all-encompassing project roster of interior design and architecture commissions, including residential, hospitality, and commercial projects. Over the years the firm has been published in numerous national and international design publications, and was named "Best of the Best" by The Robb Report. Both D'Aquino and Monaco were individually inducted into the Interior Design Hall of Fame in 2007 for their body of work.

Federico Delrosso

Federico Delrosso believes that true architecture cannot stop at the surfaces of a home, but should also develop without a break, like a Moebius strip, from the outside to the interior. In his project the exterior always finds its "natural" homolog within, but more than a simple correspondence. He enjoys working on two fundamental links: light and natural materials. Strip windows, fissures, walls dematerialized by transparency and then raw stone, rust-stained ironwork and timber with the veining clearly visible... From this minimal-naturalist approach stems the elegant lightness of the architecture and interiors designed by Federico Delrosso. In 2008 the Interiors of Montecarlo Notime restaurant has been selected as a finalist in the Hospitality/Restaurant/Casual category of "Best of Year Award".

Rocky Rockefeller

Throughout his 30-year career, Rocky has built an expansive base of experience and knowledge that ranges from leading the creative design process to meeting the technical and managerial demands of construction, project management and business. Rocky formed Rockefeller Partners in 2003 with trusted colleagues Christopher Kempel and Brian Pera. Their common purpose and shared philosophy are industriousness with a purpose and a conjoined drive to accomplish the uncommon. The firm strives to deliver excellent and diverse design services while transferring their passions for design and attention to detail onto clients and builders alike. Projects range from luxury homes of exemplary craftsmanship to complex mixed occupancies in adaptive reuse developments in the historic core of downtown Los Angeles.

Schappacher White Ltd

Schappacher White Ltd. is a New York-based architecture and interior design firm founded by Steve Schappacher and Rhea White. Their experience includes the diversity of realizing solutions for complex building projects, historic preservation, to detailed interiors and furnishings for commercial properties, corporate interiors, restaurants, nightclubs, retail, as well as a number of single family homes and residential interiors. Often this work includes custom designed furniture, fixtures and lighting for the client's specific needs.
Both their spatial solutions and product design have received the international recognition of awards, exhibits and critical acknowledgements. This multidisciplinary expertise is applied seamlessly to all levels of a project to develop creative, cost effective solutions within financial and time constraints.

Autoban

Established in 2003 by Seyhan Özdemir and Sefer Çağlar, Autoban is a design studio operating in the fields of interior design, architecture and product development.

The foundations of Autoban were laid in the Galata district of Istanbul, taking its name from the architectural landmark tower. Inspired by the chaos of the mega city, contrasts, contradictions and the co-existence of otherwise autonomous elements are all trademarks of Autoban projects and products. The Autoban studio is currently located in a grand late 19th century building in Galata's neighbouring Tunel district, a modern quarter rich with history.

Developing interior environments for retail and residential purposes, Autoban works with the leaders of a range of industries. From retail design for upscale department stores to movie theatres, cafes, restaurants and hotels; each space excels above any other for the unique living experience it offers.

Charlotte Crosland

Opening her own interior design practice in 1990, she has become renowned for creating homes and interiors that are some of the most stylish around but at the same time always comfortable places to be in. Charlotte has become a specialist in creating interiors for people with families-elegant, but not too precious. Her designs are full of bold, brave ideas that seem to suit the houses she transforms perfectly. Instead of adapting her interiors to suit furniture sourced from elsewhere, Charlotte designs her own pieces that fit seamlessly into each scheme, with clever storage and space saving solutions built in.

Charlotte Crosland Interiors, an award winning and often published interior design company, provides creative and responsive interior planning, design and construction management services. Their priority is to create environments that are immensely comfortable and effortlessly good looking.

Hector Ruiz-Velazquez

Hector Ruiz-Velazquez has a degree in architecture from the University of Virginia, USA. Hector Ruiz-Velazquez has founded his own architectural office in 1992 as a culmination of an extensive professional practice that includes architectural projects of big scale to corporate image. A characteristic of his studio is its diversity not only cultural but also regarding the professional disciplines. The studio covers projects from urbanism to graphic design, industrial and interior design, photography, as well as integral corporate images. Hector Ruiz-Velazquez' projects have been published in numerous publications all over the world since 1992 and he has been invited to lecture in different universities and public and private institutions in different parts of the world.

Slade Architecture

James Slade and Hayes Slade founded Slade Architecture in 2002, seeking to focus on architecture and design across different scales and program types. Its design approach is unique for each project but framed by a continued exploration of primary architectural concerns. Its broad definition of the project context considers any conditions affecting a specific project: program, sustainability, budget, culture, site, technology, etc. Working at the intersection of these considerations, it creates designs that are simultaneously functional and innovative.

Slade Architecture has completed a diverse range of international and domestic projects and its work has been recognized internationally with over 200 publications, exhibits and awards. Its work has been exhibited in the National Building Museum, the Museum of Modern Art and many other galleries and institutions in Europe, Asia and the United States.

René Dekker

As Head of the Residential Interiors team, René Dekker specialises in the interior design, decoration and styling of high-end and luxury residential projects. René joined SHH in 2005, after six years in the UK working for designers such as Christopher Rowley and Bill Bennette. He has been a major factor in the growth of the interiors and interior styling offer at SHH, expanding the company's housing offer from new-build, refurbishment and interior architecture projects into interior design and styling. In 2008 he was named Bathroom Designer of the Year at the KBB Awards, whilst his team's Kensington House project won the 4-star Interior Design over £100k Award at the 2008 Daily Mail UK Housebuilder Awards.

Cameron Woo

Cameron Woo, is an international award-winning interior designer and founder of his company Cameron Woo Design (CWD), located in Singapore and Sydney. Cameron's work is featured in the 2009 & 2010 Andrew Martin Interior Design Review, a publication showcasing the works of the best interior designers in the world. He has also appeared on the popular TV series, Groom My Room. In 2009 & 2010 Cameron won the Best Residential Interior Designer Award for ASIA PACIFIC, at the International Property Awards (in conjunction with Bloomberg Television). He has also worked with Asia's largest property developers such as Capitaland Residential and Kerry Properties Limited.

Brunete Fraccaroli

The renowned architect Brunete Fraccaroli, graduated at the Universidade Mackenzie in Sao Paulo and worldwide known as the colorful architecth, is a respected and requested professional for both house and commercial projects.

Among her most acknowledged projects are: The Glass Garden for "Espaco Deca"1999, which was mentioned in The New York Times, won the CREA Prize from Belo Horizonte, the 1st place at the Espaco D Award; The Glass Garage of "Casa Cor"2001 – project in honor to her father who loved cars – this area won the Solutia Design Awards 2002; the commercial project of the Glass Store in Sao Paulo, which won the contest Solutia Design Awards 2003 together with the Spanish architect Santiago Calatrava.

Helene Benhamou

Inspired by travel, art, experiences and textures, Helene Benhamou has created a style of her own - a style of elegance. Having been involved in the world of design, antiques and art for over twenty years, she has developed a fine taste for conjuring the most sophisticated and elegant spaces all around the world. Monaco, London, Morocco, New York, Paris and Singapore are just some of the locations where HB Interior has been commissioned to create the perfect living space. From the highly ornate to the sleek and modern, Helene designs extraordinary palettes that draw from all the senses to deliver spaces which are not only beautiful and elaborate but also comfortable and timeless.

Helene Benhamou lets instinct guide her work and always seems to be able to guess clients' desires. Just like a magician, she arrives in an empty space and leaves it transformed: vibrant colors, joy, light, harmony and pleasure summarize the essence of Helene Benhamou's work.

Dubbeldam Design Architects

Founded in 2002, Dubbeldam Design Architects (DDA) is a Toronto-based award winning multi-disciplinary design practice dedicated to the craft of architecture. DDA projects include architecture, interiors, landscape design, lighting and furniture design, with an emphasis on sustainable design. The firm's goal is to utilize innovative design to engage and strengthen the urban context, and to create buildings with improved energy performance. DDA's design approach focuses on renewing the city's urban fabric through modern architectural interventions, in both new construction and renovations. This philosophy, combined with a collaborative design approach with staff, clients and contractors, earned DDA the 2008 Best Emerging Practice Award by the Ontario Association of Architects.

Zhu Cheng

Education Background: Master in interior-Architecture, University of Creative Arts (United Kingdom)
Working Experience: London/Hong Kong

Ulisses Morato de Andrade

Ulisses Morato de Andrade was born in Belo Horizonte -MG, 1968. He graduated in Architecture and Urbanism in 1992 at the Methodist University Center of Minas Gerais (Centro Universitario Metodista de Minas Gerais). He has concluded a post-graduation course in Technology and Productivity in Civil Construction at the Federal University of Minas Gerais (Universidade Federal de Minas Gerais-UFMG), 2008. Since 1992, he manages Morato Arquitetura, an office that executes architecture projects, architecture of interiors and urbanism. He has projects published in books, newspapers and specialized magazines of national and international circulation. Nowadays he is the Director of Minas Gerais Department of Instituto dos Arquitetos do Brasil, IAB-MG, and editor member of the World Architecture Community portal.

II BY IV Design Associates

Founded in 1990, II BY IV Design Associates is one of Toronto's top four interior design firms and recognized in the World's Top 50. A boutique firm, we are known for cost-effective creativity and for pushing design excellence goalposts to support our clients with the best quality interiors we can. As a welcome consequence, we have been able to cultivate enduring relationships with a roster of impressive clients, many of them are global brands. To date, II BY IV has earned more than 220 awards for outstanding designs worldwide. Known for our accomplishments, we have come to be respected by our clients and industry for our innovation and attention to every detail.

Blacksheep

Cutting-edge London agency Blacksheep is in the business of creating signature spaces. Founded by designers Jo Sampson and Tim Mutton (both recently named in the prestigious Courvoisier/Observer Future 500 of upcoming business stars), Blacksheep is one of London's most creative agencies, specialists in the creation of show-stopping bars, clubs, restaurants, hotels, shops and offices. Clients include: Hermes, Park Hotels, Voyage, Gordon Ramsay, The Hilton, Diageo and Accor.

Estudio Ramos

For more than 20 years Estudio Ramos has developed a distinctive style that relies not only on a concerted vision of modernism in its designs, but also on the notion of delivering a professional and personalized service to its clients.
The close relationship that is forged with the client, plus designs that are sustainable in their environment, substantiates the firms world view of respect for humans and the environment by designing spaces that consider the impact of one on the other. These practices have enabled the firm to carry out more than 300 residential and commercial projects with success.
Established in Buenos Aires, Argentina, the firm branched out to the U.S. in 2002.
The firms principal, Juan Ignacio Ramos, brings design and practical experience, while he relies on his young associates, a son, Ignacio Ramos and daughter, Soledad Ramos plus a team of dedicated young architects to infuse the firm with innovative practices, energy and enthusiasm.

i29 interior architects

Nomination report Great Indoors Award/Interior Design Firm of the year 2008.
The fact that this firm has been nominated twice before, in the categories Concentrate & Collaborate and Serve & Facilitate, illustrates the convincing power of the projects of i29 interior architects. In the jury's eyes, this firm shows how capable it is of linking architectonic components with intensive studies of surfaces that gain maximum impact through the use of a colour or typography. This firm represents a method in which architecture and interior architecture come together in a model combination.